Atlantis Transactions in Geometry

Volume 2

Series editor

Johan Gielis, Antwerpen, Belgium

The series aims at publishing contemporary results in geometry including large parts of analysis and topology. The series will publish books of both theoretical and applied nature. Theoretical volumes will focus among other topics on submanifold theory, Riemannian and pseudo-Riemannian geometry, minimal surfaces and submanifolds in Euclidean geometry. Applications are found in biology, physics, engineering and other areas.

More information about this series at http://www.atlantis-press.com/series/15429

Johan Gielis · Paolo Emilio Ricci
Ilia Tavkhelidze
Editors

Modeling in Mathematics

Proceedings of the Second Tbilisi-Salerno Workshop on Modeling in Mathematics

ATLANTIS PRESS

Editors
Johan Gielis
Bio-Engineering Sciences
University of Antwerp
Antwerp
Belgium

Ilia Tavkhelidze
Faculty of Exact and Natural Sciences
Ivane Javakhishvili Tbilisi State University
Tbilisi
Georgia

Paolo Emilio Ricci
International Telematic University UniNettuno
Roma
Italy

Atlantis Transactions in Geometry
ISBN 978-94-6239-260-1 ISBN 978-94-6239-261-8 (eBook)
DOI 10.2991/978-94-6239-261-8

Library of Congress Control Number: 2017932628

Printed on acid-free paper

Contents

Introduction

The second volume of the Atlantis Transactions in Geometry series is a collection of papers presented at the Second Tbilisi-Salerno Workshop on Modeling in Mathematics (Tbilisi, March 16–18, 2015). This workshop resulted from a close and long standing cooperation between the Iv. Javakhishvili Tbilisi State University and the University of Salerno. The organizing committee consisted of A. Di Nola, R. Grigolia, R. Botschorishvili, J. Gielis, R. Koplatadze, T. Tadumadze, I. Tavkhelidze, M. Transirico, T. Chelidze, and R. Liparteliani. The event was cosponsored by both universities, two grants of the Shota Rustaveli National Science Foundation and the Georgian International Society of Cardiomyopathy (Georgia), The University of Antwerp, and The Simon Stevin Institute of Geometry (Belgium).

In this volume the core is on geometric modeling, expressed in geometrical, logical or analytical language, or even a combination of Boolean logic and geometry. Various chapters have a direct connection to natural phenomena, including the modeling of electromagnetic pulse propagation in biological tissues, Multi Modal Epistemic Lukasiewicz logic in immune systems, fusion in flowers, and evolution equations. Other chapters provide fundamental methods describing natural shapes and phenomena, which are fully geometric or have a close connection to geometrical or operational methods. The chapters are written in styles and language that should be accessible for a very wide audience with a geometric focus.

Gauss already stressed that the "*Geometric method will be indispensable in the early study of young people, to prevent one-sidedness and to give to the understanding a lineliness and directness, which is much less developed and—occasionally—rather jeopardized by the analytical method*". During this workshop, for his many great contributions to geometry and mathematics, the second Simon Stevin Prize for Geometry was awarded to Prof. Paolo Emilio Ricci for his fundamental contributions to mathematics in many fields, including mathematical physics, orthogonal polynomials, special functions, numerical analysis, approximation theory, and geometry. His contributions in any of these fields are fundamental, with new concepts, new methods, and new discoveries [1]. Special mention is made of his open and very inspiring style of cooperation and collaboration with colleagues worldwide, of which his coauthorship in many chapters in this volume is proof of this.

At the end of the twentieth century André Weil [2] wrote: "*Whatever the truth of the matter, mathematics in our century would not have made such impressive progress without the geometric sense of Elie Cartan, Heinz Hopf, Chern and a very few more. It seems safe to predict that such men will always be needed if mathematics is to go on as before*". It is safe to add Prof. Paolo Emilio Ricci to Weil's list of eminent geometers, to which also Prof. Bang-Yen Chen belongs as the first recipient of the Simon Stevin Prize for Geometry [3]. Simon Stevin himself was a major figure in such lists [4].

Antonio Di Nola
Johan Gielis
Revaz Grigolia
Ilia Tavkelidze

References

1. A. Cialdea., G. Dattoli, M.X. He, H.M. Srivastava, Lectures notes of the Seminario Interdisciplinare di Matematica. Volume IX. Proc. of the 4th workshop Advanced Special Functions and Solutions of PDE's on the occasion of the retirement of Paolo Emilio Ricci (2010).
2. A. Weil, S.S. Chern as Geometer and Friend, in *Shiing-Shen Chern Selected Papers*, Four Volumes, Springer-Verlag, New York, Vol. I (1978).
3. J. Gielis, First Simon Stevin Prize for Geometry. Bull. Transilvania Univ. Brasov N 15(50). Proc. Conference RIGA 2008 (2008)
4. S. Stevin, in *The Principal Works of Simon Stevin*, ed. by EJ. Dijksterhuis et al. (Zeitlinger, 1955–1966).

Fractional–Calculus–Based FDTD Algorithm for Ultra–Wideband Electromagnetic Pulse Propagation in Complex Layered Havriliak–Negami Media

Diego Caratelli, Luciano Mescia and Pietro Bia

Abstract A novel finite–difference time–domain algorithm for modeling ultra–wideband electromagnetic pulse propagation in layered multi–relaxed Havriliak–Negami media is presented. The proposed scheme is based on a general, yet computationally efficient, series representation of the fractional derivative operator associated with the permittivity function describing the frequency dispersion properties of the dielectric material. Dedicated uniaxial perfectly matched layer boundary conditions are derived and implemented in combination with the basic time–marching scheme. Moreover, a total field/scattered field formulation is adopted in order to analyze the material response under plane–wave excitation. Compared to alternative numerical methodologies available in the scientific literature, the proposed technique features a significantly enhanced robustness and accuracy which are essential for solving complex electromagnetic propagation problems typically encountered in bio–engineering applications.

1 Introduction

During the last decade, pulsed electric fields (*PEFs*) have been playing a key role in a number of new research activities in bioelectrics, a new interdisciplinary field which combine knowledge of electromagnetic principles and theory, modeling and

D. Caratelli (✉)
The Antenna Company, High Tech Campus 41, 5656 AE Eindhoven, The Netherlands
e-mail: diego.caratelli@antennacompany.com

D. Caratelli
Institute of Cybernetics, Tomsk Polytechnic University, 84/3 Sovetskaya Street, 634050 Tomsk, Russia

L. Mescia
Department of Electrical and Information Engineering, Politecnico di Bari, via E. Orabona 4, 70125 Bari, Italy
e-mail: luciano.mescia@poliba.it

P. Bia
EmTeSys, Via Beata Elia di S. Clemente, 223, 70122 Bari, Italy
e-mail: p.bia@emtesys.com

J. Gielis et al. (eds.), *Modeling in Mathematics*,
Atlantis Transactions in Geometry 2, DOI 10.2991/978-94-6239-261-8_1

simulations, physics, material science, cell biology, and medicine. The main goal of this discipline is the study of the interaction between electromagnetic fields and biological tissues aimed at the investigation of nanopulse bioeffects for human safety as well as for at use of ultrashort pulses in biomedical and biotechnological applications [1–5].

Current applications of *PEF* are primarily based on reversible or irreversible electroporation, that is the process by which the permeability to drugs, molecules, and genetic material of the plasma membrane of a biological cell is affected. This phenomenon can result either permanent permeabilization of cancer cells or the destabilization of the cell membranes and intracellular components useful to trigger cellular mechanism leading to cellular death [6–8]. As a result, electroporation based therapies and treatments can be used to achieve selective killing of cancer cells, tissue ablation, gene therapy and *DNA* based vaccination. In particular, in the cancer therapy *PEFs* have been defined for the treatment of easily accessible cutaneous and subcutaneous tumor nodules, prostate cancer, fibrosarcoma. This approach has demonstrated remarkable potential for the treatment of solid tumors without hyperthermia or delivering drugs or genes [3–5, 9]. Other therapeutic applications include coronary and peripheral vascular disease, activating platelets, plasmids transfection, immune responses enhancement, tissue imaging [3]. Moreover, the application of this new technology could have a great impact on molecular biology, by promoting the understanding of molecular mechanisms of cells.

The efficiency of *PEF* treatment depends on the electric field distribution within the treated tissue. In fact, the cellular death mechanism is strongly affected by pulse parameters such as amplitude, duration, number of pulses and repetition frequency. However, *PEF* exposure could result in side effects such as tissue damage, conformational changes in macromolecules, alteration of the biochemical reaction rates, membrane characteristics and temperature levels [10]. Therefore, in order to predict the effect of exposures and assess possible outcomes it is very important to know the local electric field distribution inside the exposed tissue. In many applicative cases, the electromagnetic field cannot be easily measured. To this aim, theoretical models are invaluable tools to better understand the involved mechanisms as well as to evaluate and optimize treatment modalities and to develop disease–specific or even patient–specific protocols.

The electric field distribution excited in biological media mainly depends on the electric properties of tissues. But the lack of data and accurate models, over broad frequency ranges, has been so far an obstacle for both theoretical and experimental studies. In fact, the complexity of the structure and composition of biological matter produces anomalies in the dynamic dielectric properties resulting in a strong dispersion of dielectric susceptibility. This dispersion, can be explained by considering that the disordered nature and microstructure of the systems yield a multiple relaxation times. As a result, the time-domain response is generally non–symmetric and markedly different from that of dielectric media modeled by the simple Debye equation. Therefore, it is important to define empirical models for each organ, in a wide frequency range, for studies regarding the interaction between electromagnetic wave and biological bodies, especially for *PEF* excitation.

The dielectric properties of biological tissues result from the interaction of electromagnetic energy with the tissue constituents at the cellular and molecular level. This process is strongly affected by the bound water content. The frequency variation of the dielectric properties of tissues with high water content can be easily described by the Debye relation. This kind of response is obtained for an assembly of identical dipoles yielding a loss of energy proportional to the electric field frequency. However, it is well known that the dielectric properties of many biological materials display α, $-\beta$, $-\gamma$ and $-\delta$ dispersion types attributed to a different polarization mechanism [11]. As a consequence, an accurate representation of the experimental dielectric response in frequency domain usually cannot be described by a simple exponential expression with a single relaxation time (Debye model). To this end, a number of empirical relationships including Cole-Cole (*CC*), Cole-Davidson (*CD*) and Havriliak-Negami (*HN*) equations have been proposed in order to fit such types of dielectric spectra. The Cole–Cole model is generally chosen to describe the relative complex permittivity of many types of biological tissues over wide frequency ranges. However, *HN* representation includes both *CC* and *CD* models and provides an extended model flexibility enabling a better parametrization of the arbitrary dispersive media properties.

The accurate modeling of electromagnetic field propagation in the mentioned dispersive materials is essential for gaining a deeper insight into the physical mechanisms affecting the interaction between pulsed electric fields (*PEFs*) and biological media. Finite–difference time–domain (*FDTD*) method has been widely used in electromagnetic modeling due to its straightforward implementation and ability to model a broad range of exposure conditions [12, 13]. Since *CC*, *CD*, and *HN* dispersion functions include fractional powers of the angular frequency, suitable mathematical models adequately describing the response of such complex dispersive media have to be embedded in the core of the *FDTD* algorithm [14–19].

Recently, a novel *FDTD* methodology for modeling Havriliak–Negami (*HN*) media has been presented by the authors in [1]. Said formulation is based on the optimal truncation of the binomial series related to the *HN* fractional derivative operator, in accordance with the Riemann–Liouville theory. Typically, the aforementioned truncated series approach provides a very good approximation of both real and imaginary parts of *HN* permittivity functions, and numerous test cases discussed in [1, 2] demonstrate that the proposed scheme is reliable and accurate over broad frequency ranges. However, further investigations have highlighted some numerical inaccuracies wherein specific *HN* materials have to be modeled over ultra–wide bands, at angular frequencies $\omega \gg 1/\tau$, τ denoting the characteristic relaxation time of the medium [20, 21]. In order to overcome these limitations, the authors have extended the previously presented *FDTD* scheme by implementing a more general series representation of the *HN* fractional derivative operator in order to account for multiple relaxation times and ohmic losses occurring in the considered biological medium. The enhanced accuracy of the modified *FDTD* procedure has been assessed by several test cases involving complex stratified *HN* media, and compared against a fully analytical modeling approach.

2 Mathematical Formulation

The dielectric properties of biological tissues are determined by the interaction of the electromagnetic energy with the tissue constituents at cellular and molecular level. As a result, the permittivity and electrical conductivity vary from tissue to tissue and depend on the working frequency. In order to accurately model the electromagnetic wave propagation over broad frequency ranges, suitable analytical models of the dielectric properties are needed. In particular, the macroscopic characteristics of general dispersive media can be modeled by using the following multi–relaxation *HN* relationship:

$$\varepsilon_r(\omega) = \varepsilon_{r_\infty} + \sum_{l=1}^{N} \frac{\varepsilon_{r_{s,l}} - \varepsilon_{r_\infty}}{\left[1 + (j\omega\tau_l)^{\alpha_l}\right]^{\beta_l}} - j\frac{\sigma}{\omega\varepsilon_0}, \tag{1}$$

where σ is the static ionic conductivity, ε_{r_∞} is the asymptotic relative permittivity for $\omega\tau \to +\infty$, τ_l and $\varepsilon_{r_{s,l}}$ denote the l–th relaxation time and static relative permittivity (for $\omega\tau \to 0^+$) respectively, and $0 < \alpha_l,\ \beta_l < 1$ are heuristically derived model fitting parameters for $l = 1, 2, \ldots, N$, with N being the maximum number of relaxation processes occurring in the considered media. In (1), as usual, $\omega = 2\pi f$ is the angular frequency, and ε_0 denotes the permittivity of free space.

In the proposed *FDTD* scheme, the following approximated expansion is adopted in place of the truncated binomial series presented in [1]:

$$\underbrace{\left[1 + (j\omega\tau_l)^{\alpha_l}\right]^{\beta_l}}_{F(j\omega)} \simeq \underbrace{\sum_{n=0}^{K_l} \chi_{n,l}\,(j\omega\tau_l)^{\zeta_{n,l}}}_{F^a(j\omega)}, \tag{2}$$

where the parameters $\zeta_{n,l}$ and $\chi_{n,l}$ are assumed to satisfy the inequalities $|\chi_{n,l}| < a$ and $0 < \zeta_{n,l} < b$, with a and b denoting assigned positive real numbers.

Let $K_{\max}$ be the maximum expansion order in (2), and δ be a given small positive threshold to be used for controlling the accuracy of the approximation. In this way, the parameters K_l, $\zeta_{n,l}$, and $\chi_{n,l}$ ($l = 1, 2, \ldots, N$) can be evaluated as follows:

1. Initialize $K_l = 1$, and set the working frequency range $\omega_{\min} < \omega < \omega_{\max}$;
2. Calculate $\chi_{n,l}$ and $\zeta_{n,l}$ by using the Nelder–Mead algorithm [22];
3. Evaluate the relative error function

$$e_r = \sqrt{\frac{\int_{\omega_{\min}}^{\omega_{\max}} |F(j\omega) - F^a(j\omega)|^2 d\omega}{\int_{\omega_{\min}}^{\omega_{\max}} |F(j\omega)|^2 d\omega}}. \tag{3}$$

4. If $e_r \le \delta$ or $K_l = K_{\max}$, the algorithm stops, else update the expansions orders as $K_l = K_l + 1$, and go to step 2.

The evaluation of the electromagnetic field distribution excited within the multi–layered dielectric material under analysis can be then performed by using the *FDTD* scheme proposed in [1], that has been here further extended in order to take the ohmic losses as well as the multi–relaxation response of the medium into account.

Let us consider a non-magnetic dispersive medium with complex relative permittivity described by (1). Under such assumption, the differential version of the Ampere's law in time domain, within said material, can be written as:

$$\nabla \times \mathbf{H} = \varepsilon_0 \varepsilon_{r_\infty} \partial_t \mathbf{E} + \sigma \mathbf{E} + \sum_{l=1}^{N} \mathbf{J}_l, \tag{4}$$

with ∂_t denoting the partial derivative operator with respect to time, and where the auxiliary displacement current density terms $\mathbf{J}_l$ ($l = 1, 2, \ldots, N$) have been introduced. It is straightforward to find out that the k–th term ($1 \leq k \leq N$) is such to satisfy the equation:

$$\mathrm{D}_t^{\alpha_k, \beta_k} \mathbf{J}_k = \varepsilon_0 \Delta \varepsilon_{r_k} \partial_t \mathbf{E}, \tag{5}$$

involving the fractional derivative operator $\mathrm{D}_t^{\alpha_k,\beta_k} = \mathcal{F}^{-1}\left\{\left[1 + (j\omega\tau_k)^{\alpha_k}\right]^{\beta_k}\right\} = \left(1 + \tau_k^{\alpha_k} \mathrm{D}_t^{\alpha_k}\right)^{\beta_k}$ as defined in [1]. In (5), for the sake of brevity, the scalar quantities $\Delta\varepsilon_{r_k} = \varepsilon_{r_{s,k}} - \varepsilon_{r_\infty}$ have been used.

Upon substituting (5) in (4), and applying a second–order accurate finite–difference scheme, one readily obtains, at the time instant $t = m\Delta t$:

$$(\nabla \times \mathbf{H})|^m - \frac{\varepsilon_{r_\infty}}{\Delta \varepsilon_{r_k}} \left(\mathrm{D}_t^{\alpha_k,\beta_k} \mathbf{J}_k\right)\Big|^m = \sum_{l=1}^{N} \mathbf{J}_l|^m + \sigma\ \mathbf{E}|^m, \tag{6}$$

where the vector terms appearing on the right–hand side of the equation are evaluated by means of the semi–implicit approximation:

$$\left.\begin{Bmatrix}\mathbf{J}_l \\ \mathbf{E}\end{Bmatrix}\right|^m = \frac{1}{2}\left(\left.\begin{Bmatrix}\mathbf{J}_l \\ \mathbf{E}\end{Bmatrix}\right|^{m-\frac{1}{2}} + \left.\begin{Bmatrix}\mathbf{J}_l \\ \mathbf{E}\end{Bmatrix}\right|^{m+\frac{1}{2}}\right). \tag{7}$$

In a similar way, from Eq. (5) it follows that:

$$\mathbf{E}|^{m+\frac{1}{2}} = \mathbf{E}|^{m-\frac{1}{2}} + \frac{\Delta t}{\varepsilon_0 \Delta \varepsilon_{r_k}} \left(\mathrm{D}_t^{\alpha_k,\beta_k} \mathbf{J}_k\right)\Big|^m, \tag{8}$$

where:

$$\mathrm{D}_t^{\alpha_k,\beta_k} \mathbf{J}_k \simeq \sum_{n=0}^{K_k} \chi_{n,k} \tau_k^{\zeta_{n,k}} \mathrm{D}_t^{\zeta_{n,k}} \mathbf{J}_k, \tag{9}$$

Let $\nu_{n,k}$ be the integer number such that $\nu_{n,k} - 1 \leqslant \zeta_{n,k} \leqslant \nu_{n,k}$. So, applying the Riemann–Liouville definition of fractional derivative, the following equation is derived:

$$\mathrm{D}_t^{\zeta_{n,k}} \mathbf{J}_k(t) = \frac{\mathrm{d}^{\nu_{n,k}}}{\mathrm{d}t^{\nu_{n,k}}} \int_0^t \frac{(t-u)^{\nu_{n,k}-\zeta_{n,k}-1}}{\Gamma\left(\nu_{n,k}-\zeta_{n,k}\right)} \mathbf{J}_k(u)\,\mathrm{d}u. \tag{10}$$

By setting:

$$\mathbf{I}_{n,k}(t) = \int_0^t (t-u)^{\nu_{n,k}-\zeta_{n,k}-1} \, \mathbf{J}_k(u)\,\mathrm{d}u, \tag{11}$$

and applying a central finite difference approximation with time step Δt, one can readily obtain, at the general time instant $t = m\Delta t$:

$$\begin{aligned} \mathbf{I}_{n,k}\big|^m &\simeq \sum_{p=0}^{m-1} \mathbf{J}_k|^{m-p-\frac{1}{2}} \int_{p\Delta t}^{(p+1)\Delta t} u^{\nu_{n,k}-\zeta_{n,k}-1}\mathrm{d}u \\ &= \frac{\Delta t^{\nu_{n,k}-\zeta_{n,k}}}{\nu_{n,k}-\zeta_{n,k}} \sum_{p=0}^{m-1} \left[(p+1)^{\nu_{n,k}-\zeta_{n,k}} - p^{\nu_{n,k}-\zeta_{n,k}}\right] \mathbf{J}_k|^{m-p-\frac{1}{2}}. \end{aligned} \tag{12}$$

By using the following expansion [23]:

$$(p+1)^{\nu_{n,k}-\zeta_{n,k}} - p^{\nu_{n,k}-\zeta_{n,k}} \simeq \sum_{q=1}^{Q_{n,k}} a_{n,k,q} e^{-b_{n,k,q} p}, \tag{13}$$

with the order $Q_{n,k}$ and the coefficients $a_{n,k,q}$ and $b_{n,k,q}$ being suitably chosen in order to minimize the mean square error, and upon setting:

$$\mathbf{\Psi}_{n,k,q}\big|^m = \sum_{p=0}^{m-1} a_{n,k,q} e^{-b_{n,k,q} p} \, \mathbf{J}_k|^{m-p-\frac{1}{2}} = a_{n,k,q} \, \mathbf{J}_k|^{m-\frac{1}{2}} + e^{-b_{n,k,q}} \, \mathbf{\Psi}_{n,k,q}\big|^{m-1}, \tag{14}$$

equation (12) can be rewritten as:

$$\mathbf{I}_{n,k}\big|^m \simeq \frac{\Delta t^{\nu_{n,k}-\zeta_{n,k}}}{\nu_{n,k}-\zeta_{n,k}} \left(S_{n,k} \, \mathbf{J}_k|^{m-\frac{1}{2}} + \sum_{q=1}^{Q_{n,k}} e^{-b_{n,k,q}} \, \mathbf{\Psi}_{n,k,q}\big|^{m-1} \right), \tag{15}$$

where:

$$S_{n,k} = \sum_{q=1}^{Q_{n,k}} a_{n,k,q}. \tag{16}$$

In (10), the time derivative of $\mathbf{I}_{n,k}$ calculated at the time instant $t = m\Delta t$ can be approximated by means of the expression:

$$\left.\frac{\mathrm{d}^{\nu_{n,k}}\mathbf{I}_{n,k}}{\mathrm{d}t^{\nu_{n,k}}}\right|^{m} \simeq \frac{1}{(\Delta t)^{\nu_{n,k}}}\sum_{p=0}^{\nu_{n,k}}(-1)^p\binom{\nu_{n,k}}{p}\mathbf{I}_{n,k}\Big|^{m-p+1}$$
$$= \frac{\Delta t^{-\zeta_{n,k}}}{\nu_{n,k}-\zeta_{n,k}}\left\{S_{n,k}\left[\mathbf{J}_k|^{m+\frac{1}{2}}+\sum_{p=1}^{\nu_{n,k}}(-1)^p\binom{\nu_{n,k}}{p}\mathbf{J}_k|^{m-p+\frac{1}{2}}\right]\right.$$
$$\left.+\sum_{p=1}^{\nu_{n,k}}(-1)^p\binom{\nu_{n,k}}{p}\left(\sum_{q=1}^{Q_{n,k}}e^{-b_{n,k,q}}\,\mathbf{\Psi}_{n,k,q}\Big|^{m-p}\right)\right\}. \tag{17}$$

In this way, after some algebra, it is not difficult to show that:

$$\left(\mathrm{D}_t^{\alpha_k,\beta_k}\mathbf{J}_k\right)\Big|^{m} \simeq \sum_{n=0}^{K_k}\frac{\chi_{n,k}}{(\nu_{n,k}-\zeta_{n,k})!}\left(\frac{\tau_k}{\Delta t}\right)^{\zeta_{n,k}}\sum_{p=0}^{\nu_{n,k}}\left[A_{n,k,p}\,\mathbf{J}_k|^{m-p+\frac{1}{2}}\right.$$
$$\left.+\sum_{q=1}^{Q_{n,k}}B_{n,k,p,q}\,\mathbf{\Psi}_{n,k,q}\Big|^{m-p}\right], \tag{18}$$

with:

$$\begin{Bmatrix}A_{n,k,p}\\B_{n,k,p,q}\end{Bmatrix} = (-1)^p\binom{\nu_{n,k}}{p}\begin{Bmatrix}S_{n,k}\\e^{-b_{n,k,q}}\end{Bmatrix}, \tag{19}$$

Finally, by combining (6) with (7), (8), and (18), one can readily obtain:

$$\left[\left(\varepsilon_{r_\infty}+\frac{\sigma\Delta t}{2\varepsilon_0}\right)\frac{C_k}{\Delta\varepsilon_{r_k}}+\frac{1}{2}\right]\mathbf{J}_k|^{m+\frac{1}{2}}+\frac{1}{2}\sum_{l=1,l\neq k}^{N}\mathbf{J}_l|^{m+\frac{1}{2}} =$$
$$= (\nabla\times\mathbf{H})|^{m}-\sigma\mathbf{E}|^{m-\frac{1}{2}}-\frac{1}{2}\sum_{l=1}^{N}\mathbf{J}_l|^{m-\frac{1}{2}}-\left(\varepsilon_{r_\infty}+\frac{\sigma\Delta t}{2\varepsilon_0}\right)$$
$$\cdot\frac{1}{\Delta\varepsilon_{r_k}}\sum_{n=0}^{K_k}\frac{\chi_{n,k}}{(\nu_{n,k}-\zeta_{n,k})!}\left(\frac{\tau_k}{\Delta t}\right)^{\zeta_{n,k}}\left[\sum_{p=1}^{\nu_{n,k}}A_{n,k,p}\,\mathbf{J}_k|^{m-p+\frac{1}{2}}\right.$$
$$\left.+\sum_{p=0}^{\nu_{n,k}}\sum_{q=1}^{Q_{n,k}}B_{n,k,p,q}\,\mathbf{\Psi}_{n,k,q}\Big|^{m-p}\right] = \boldsymbol{\eta}_k\Big|^{m}, \tag{20}$$

where:

$$C_k = \sum_{n=0}^{K_k}A_{n,k,0}\frac{\chi_{n,k}}{\nu_{n,k}-\zeta_{n,k}}\left(\frac{\tau_k}{\Delta t}\right)^{\zeta_{n,k}}, \tag{21}$$

for $k = 1, 2, \ldots, N$. It is apparent from (20) that in the presented formulation, contrary to the methodology in [1], the evaluation of the displacement current density entails solving a symmetric system of N linear equations, this reflecting the multi–relaxation characteristics of the dielectric material under analysis. As a matter of fact, Eq. (19) can be recast in the more compact matrix form:

$$\underbrace{\left[\frac{1}{2}\underline{\mathbf{U}}+\left(\varepsilon_{r_\infty}+\frac{\sigma\,\Delta t}{2\varepsilon_0}\right)\underline{\mathbf{D}}\right]}_{\underline{\mathbf{T}}}\cdot\mathbf{J}|^{m+\frac{1}{2}}=\boldsymbol{\eta}|^{m}\,, \tag{22}$$

with $\underline{\mathbf{U}}$ being the unit matrix of order N, and $\underline{\mathbf{D}} = \operatorname{diag}\{D_1, D_2, \ldots, D_N\}$ the diagonal matrix with nonzero entries $D_k = C_k/\Delta\varepsilon_{r_k}$. In (22), $\mathbf{J}|^{m+\frac{1}{2}}$ denotes the vector of the unknown current densities at the time instant $t = \left(m+\frac{1}{2}\right)\Delta t$, namely:

$$\mathbf{J}|^{m+\frac{1}{2}} = \begin{bmatrix} \mathbf{J}_1|^{m+\frac{1}{2}} \\ \mathbf{J}_2|^{m+\frac{1}{2}} \\ \vdots \\ \mathbf{J}_N|^{m+\frac{1}{2}} \end{bmatrix}. \tag{23}$$

Similarly, the column vector $\boldsymbol{\eta}|^m$ is built up by arraying the auxiliary electromagnetic field quantities $\boldsymbol{\eta}_k\big|^m$ $(k = 1, 2, \ldots, N)$ appearing on the right–hand side of (20). It is worth noting that the inverse of the coefficient matrix $\underline{\mathbf{T}}$ of the linear system (22) can be conveniently computed only one time before the time–marching scheme is initiated. In this way, in comparison to the *FDTD* procedure described in [1], the algorithmic implementation of the technique proposed in this research study actually results in a reduced additional computational cost of $O\left(N^2\right)$ floating–point operations useful to determine the solution of (20) as:

$$\mathbf{J}|^{m+\frac{1}{2}} = \underline{\mathbf{T}}^{-1}\cdot\boldsymbol{\eta}|^{m}\,. \tag{24}$$

Once the current density terms $\mathbf{J}_l|^{m+\frac{1}{2}}$ $(l = 1, 2, \ldots, N)$ are evaluated, the electric field distribution within the considered dielectric medium can be derived from (6) as:

$$\mathbf{E}|^{m+\frac{1}{2}} = \frac{2\varepsilon_0\varepsilon_{r_\infty}-\sigma\,\Delta t}{2\varepsilon_0\varepsilon_{r_\infty}+\sigma\,\Delta t}\,\mathbf{E}|^{m-\frac{1}{2}} + \frac{2\Delta t}{2\varepsilon_0\varepsilon_{r_\infty}+\sigma\,\Delta t}\left[(\nabla\times\mathbf{H})|^{m}\right.$$
$$\left.-\frac{1}{2}\sum_{l=1}^{N}\left(\mathbf{J}_l|^{m-\frac{1}{2}}+\mathbf{J}_l|^{m+\frac{1}{2}}\right)\right], \tag{25}$$

where judicious use of (7) has been made. Finally, by carrying out a second–order accurate finite–difference approximation of the Faraday's law in the time domain, the following update equation for the magnetic field is readily obtained:

$$\mathbf{H}|^{m+1} = \mathbf{H}|^{m} - \frac{\Delta t}{\mu_0} \left(\nabla \times \mathbf{E}\right)|^{m+\frac{1}{2}}, \tag{26}$$

with μ_0 denoting the magnetic permeability of free space.

In order to truncate the *FDTD* computational domain and solve electromagnetic problems with open boundaries, dedicated uniaxial perfectly matched layer (*UPML*) conditions [24] have to be derived and implemented numerically accounting for the electrical conductivity and the multi–relaxation characteristics of the *HN* medium under analysis. To this end, let us first introduce the auxiliary electric field vector $\mathbf{e}$ as:

$$\mathbf{e} = \left(\kappa_z + \frac{\sigma_z}{j\omega\varepsilon_0}\right)\mathbf{E}, \tag{27}$$

with κ_z, σ_z denoting the *UPML* material parameters in accordance with the complex coordinate stretching approach [12]:

$$\kappa_z(z) = \begin{cases} 1, & z \notin \text{UPML} \\ 1 + (\kappa_{\text{MAX}} - 1)\left(\dfrac{z - z^*_{\text{UPML}}}{d_{\text{UPML}}}\right)^m, & z \in \text{UPML} \end{cases} \tag{28}$$

$$\sigma_z(z) = \begin{cases} 0, & z \notin \text{UPML} \\ \sigma_{\text{MAX}}\left(\dfrac{z - z^*_{\text{UPML}}}{d_{\text{UPML}}}\right)^m, & z \in \text{UPML} \end{cases} \tag{29}$$

In (28)–(29), z^*_{UPML} and d_{UPML} denote, respectively, the coordinate of the *UPML* interface and the relevant thickness. In particular, σ_{MAX} is the maximum value assumed by the conductivity at the truncation of the *UPML* region $\left(z = z^*_{\text{UPML}} + d_{\text{UPML}}\right)$ and κ_{MAX} is a real coefficient which is selected heuristically in order to enhance the absorption of electromagnetic waves within the *UPML* region and, in this way, minimize the spurious reflection level in the solution domain.

Multiplying both sides of (27) by $j\omega$ and transforming into the time domain immediately yields:

$$\partial_t \mathbf{e} = \kappa_z \partial_t \mathbf{E} + \frac{\sigma_z}{\varepsilon_0}\mathbf{E}. \tag{30}$$

In this way, it is not difficult to find out that the Ampere's law can be written, within the *UPML* region, as:

$$\nabla \times \mathbf{H} = \varepsilon_0 \varepsilon_{r_\infty} \partial_t \mathbf{e} + \sigma \mathbf{e} + \sum_{l=1}^{N} \mathbf{j}_l, \tag{31}$$

where the l–th displacement current density term satisfies the fractional derivative equation:

$$\mathrm{D}_t^{\alpha_l,\beta_l} \mathbf{j}_l = \varepsilon_0 \Delta\varepsilon_{r_l} \partial_t \mathbf{e}. \tag{32}$$

The discretization of the considered equations on the Yee lattice can be conveniently carried out by adopting the usual leapfrog scheme in time, wherein the loss terms are averaged according to the semi–implicit approximation [see (7)]. Overall, as it can be easily figured out by comparison with (4) and (5), this leads to time–stepping expressions which are formally equivalent to (20) and (25) with the stretched vectors $\mathbf{e}$ and $\mathbf{j}_l$ replacing the quantities $\mathbf{E}$ and $\mathbf{J}_l$, respectively, for $l = 1, 2, \ldots, N$. On the other hand, the electric field update equation directly follows from (6) as:

$$\mathbf{E}|^{m+\frac{1}{2}} = \frac{2\varepsilon_0\kappa_z - \sigma_z\Delta t}{2\varepsilon_0\kappa_z + \sigma_z\Delta t}\,\mathbf{E}|^{m-\frac{1}{2}} + \frac{2\varepsilon_0}{2\varepsilon_0\kappa_z + \sigma_z\Delta t}\left(\mathbf{e}|^{m+\frac{1}{2}} - \mathbf{e}|^{m+\frac{1}{2}}\right). \qquad (33)$$

Finally, by Fourier–transforming the Faraday's law $\nabla \times \mathbf{E} = -\mu_0 \left(j\omega\kappa_z + \sigma_z/\varepsilon_0\right)\mathbf{H}$ and discretizing at the time instance $t = \left(m + \frac{1}{2}\right)\Delta t$, one can readily derive the time–stepping expression for the magnetic field within the *UPML* termination:

$$\mathbf{H}|^{m+1} = \frac{2\varepsilon_0\kappa_z - \sigma_z\Delta t}{2\varepsilon_0\kappa_z + \sigma_z\Delta t}\,\mathbf{H}|^{m} - \frac{2Y_0^2\Delta t}{2\varepsilon_0\kappa_z + \sigma_z\Delta t}\left(\nabla \times \mathbf{E}\right)|^{m+\frac{1}{2}}, \qquad (34)$$

with $Y_0 = \sqrt{\varepsilon_0/\mu_0}$ being the wave admittance in free space.

3 Numerical Results

The validation of the developed numerical procedure has been carried out by three different one–dimensional test cases involving multi–layered *HN* dielectric materials under plane–wave excitation. To this end, the conventional total field/scattered field (*TF/SF*) formulation detailed in [12] has been implemented using a sinusoidally time–modulated Gaussian pulse source with the same characteristic parameters reported in [1]. In all cases, the truncates series approximation of the *HN* fractional derivative operator is performed by means of the fitting algorithm described in the previous section. In particular, the parameters a and b have been set heuristically to 10 and 1, respectively, based on the features of the search domain, such that $0 \leqslant \alpha_l, \beta_l \leqslant 1$, with $\tau_l > 0$ for every index l. It has been noted that, for $b > 1$ the *FDTD* scheme is potentially prone to numerical instability. On the other hand, an exhaustive numerical verification has been led to the conclusion that, by assuming $b = 1$, a very accurate series approximation of (1) is achieved for each *HN* parameter. In all computations, the error threshold $\delta = 10^{-6}$ and maximal model order $K_{\max} = 6$ have been assumed.

The first test case is relevant to a single–layered *HN* dielectric slab in air. The material parameters are $\alpha = 0.9$, $\beta = 0.3$, $\varepsilon_{r_\infty} = 4$, $\Delta\varepsilon_r = 88$, $\sigma = 0$, and $\tau = 140\,\text{ps}$, the thickness of the slab being $d = 10\,\text{mm}$. Figure 1 shows, in logarithmic scale, the relative fitting error e_r as a function of α and β in comparison to the one achieved by means of the truncated series expansion illustrated in [1]. As it can be noticed, the

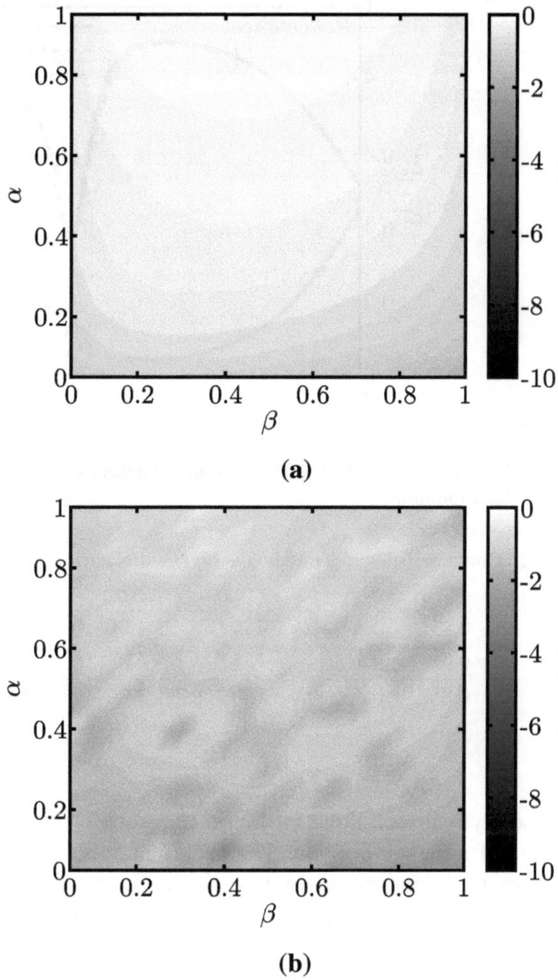

Fig. 1 Error function versus the parameters α and β of a *HN* material with single relaxation time $\tau = 140\,\text{ps}$ when modeled with **a** the approach presented in [1], and **b** the new methodology detailed in this study

newly introduced fitting procedure provides a significantly more accurate representation of the *HN* fractional derivative operator with a maximum error value, over the complete problem space, of about 4.2% contrary to the methodology in [1] which, on the other hand, can result in a maximal error of about 45.0%, this being due to the fact that, for the selected *HN* parameters, the truncated binomial series, even if it is never divergent, provides a poor approximation for $\omega\tau > 1$. In this context, it is worth stressing that the selection of the optimal number of series terms in (2) has been made by applying the developed Nelder–Mead–based algorithm (see Sect. 2), while enforcing the positivity of the approximated *HN* permittivity function in order to ensure the physical consistency of the model. Figure 2 shows the frequency–domain behavior of the reflectance and transmittance of the considered dielectric slab as evaluated by using the new *FDTD* procedure, as well as the scheme in [1], and a rigorous

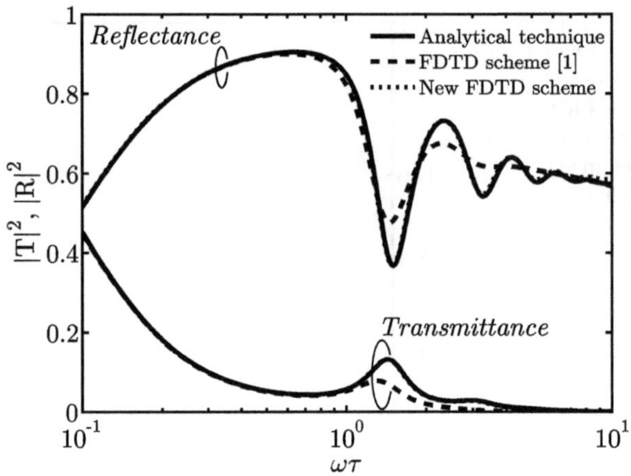

Fig. 2 Reflectance and transmittance versus frequency of a single–layered *HN* dielectric slab with single relaxation time

Table 1 Characteristics of a three–layered *HN* dielectric slab with single relaxation time

Medium	α	β	τ (ps)	$\Delta\varepsilon_r$	σ (S/m)	$\varepsilon_{r\infty}$	d (mm)
1	0.85	0.3	135	48	0	4	10
2	0.9	0.5	155	65	0.01	2.5	5
3	0.8	0.8	140	33	0.06	2.5	15

fully analytical technique based on the transfer matrix approach for layered lossy media in frequency domain. The excellent agreement with the analytical technique validates the proposed methodology. On the other hand, the inappropriateness of the original *FDTD* scheme to model the considered *HN* medium at high frequencies is apparent.

The second test case consists of a three–layered *HN* dielectric slab in air, featuring the geometrical and electrical parameters listed in Table 1. As it can be observed in Fig. 3, even in this case, the new *FDTD* formulation allows a more accurate electromagnetic modeling in comparison to the original scheme in [1], the difference being more noticeable at higher frequencies.

The third test case is the more general one since it is relevant to the electromagnetic characterization of a three–layered dispersive dielectric slab over an ultra–wide frequency band spanning from $f_{\min} = 100$ MHz to $f_{\max} = 10$ GHz. Each layer of the slab is assumed to be made out of a *HN* medium having the geometrical and electrical parameters listed in Table 2. In order to assess the reliability and robustness of the developed numerical code, very different relaxation times $\tau_{i,j}$ are considered across the various material layers. As a matter of fact, with the specific choice of parameters in Table 2, the normalized frequency $2\pi f_{\max}\tau_{i,j}$ ranges from about 0.5 to about 430. The frequency–domain behavior of reflectance and transmittance featured by the

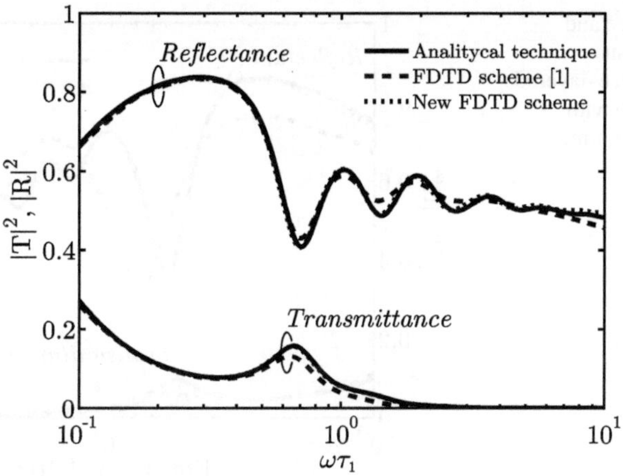

Fig. 3 Reflectance and transmittance versus frequency of a three–layered *HN* dielectric slab with single relaxation time (see Table 1)

Table 2 Characteristics of a three–layered *HN* dielectric slab with multiple relaxation times

Medium	$\alpha_{1,2}$	$\beta_{1,2}$	$\tau_{1,2}$	$\Delta\varepsilon_{r1,2}$	σ(S/m)	$\varepsilon_{r\infty}$	d(mm)
1	0.93	0.5	8 ps	37	0	4	8
	0.92	0.57	6.8 ns	179			
2	0.92	0.6	8.3 ps	2.3	0.01	2.5	10
	0.91	0.35	2.3 ns	79.2			
3	0.91	0.7	13.8 ps	8.2	0.06	2.5	12
	0.7	0.3	6.4 ns	130			

considered dielectric slab as evaluated by using the proposed *FDTD* procedure, as well as the scheme in [1], and the aforementioned analytical transfer–matrix–based approach is reported in Fig. 4. It is apparent from the obtained results that the original *FDTD* technique is not suitable for handling such complex material characteristics. Further numerical investigations have shown that the conventional binomial series approximation in [1] becomes inaccurate when the normalized frequency $\omega\tau$ is typically larger than 10, as it occurs in biological media displaying different dispersion mechanisms. On the other hand, the methodology illustrated in Sect. 2 is in excellent agreement with theory over the complete operating frequency range.

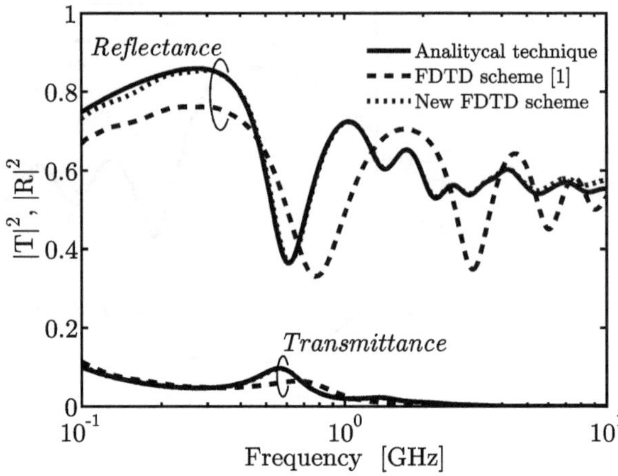

Fig. 4 Reflectance and transmittance versus frequency of a three–layered *HN* dielectric slab with multiple relaxation times (see Table 2)

4 Conclusion

A novel *FDTD* technique based on the fractional derivative operator theory has been presented. The developed numerical approach allows analyzing short–pulse wave propagation in general layered *HN* dispersive materials characterized by multiple relaxation times. Dedicated *UPML* boundary conditions have been derived and implemented in combination with the basic time–marching scheme.

The soundness and reliability of the modified *FDTD* procedure has been assessed by several test cases. The obtained results show a good agreement with theory, and confirm the enhancement of the numerical accuracy of the methodology in comparison with the previously one published by the authors, especially when the ultra–wideband characterization of multi–relaxed media is to be carried out.

The presented technique provides a general–purpose tool useful to address complex electromagnetic problems in the field of bio–engineering. Its application to the study of temperature elevation in biological tissues subject to *PEF* excitation, as well as the investigation of electroporation processes in cells exposed to high–voltage electric pulses is currently ongoing.

Acknowledgements This study has been partly carried out in the framework of the research and development program running at *The Antenna Company*. For further information, please visit the Web site: `http://www.antennacompany.com/`.

References

1. L. Mescia, P. Bia, D. Caratelli, Fractional derivative based FDTD modeling of transient wave propagation in Havriliak-Negami media. IEEE Trans. Microw. Theory Tech. **62**, 1920–1929 (2014)

2. P. Bia, D. Caratelli, L. Mescia, R. Cicchetti, G. Maione, F. Prudenzano, A novel FDTD formulation based on fractional derivatives for dispersive Havriliak-Negami media. Signal Process. **107**, 312–318 (2015)
3. S.J. Beebe, Bioelectrics in basic science and medicine: impact of electric fields on cellular structures and functions. J. Nanomed. Nanotechol. **4**, 1000163 (2013)
4. S. Corovic, I. Lackovic, P. Sustaric, T. Sustar, T. Rodic, D. Miklavcic, Modeling of electric field distribution in tissues during electroporation. BioMed. Eng. **12**, 2–27 (2013)
5. A.T. Esser, K.C. Smith, T.R. Gowrishankar, J.C. Weaver, Towards solid tumor treatment by nanosecond pulsed electric fields. Technol. Cancer Res. Treat. **8**, 289–306 (2009)
6. J.C. Weaver, Electroporation of cells and tissues. IEEE Trans. Plasma Sci. **88**, 28–33 (2000)
7. J. Gehl, Electroporation: theory and methods, perspectives for drug delivery, gene therapy and research. Acta Physiol. Scand. **177**, 437–447 (2003)
8. J. Teissié, J.M. Escoffre, M.P. Rols, M. Golzio, Time dependence of electric field effects on cell membranes. A review for a critical selection of pulse duration for therapeutical applications. Radiol. Oncol. **42**, 196–206 (2008)
9. R. Nuccitelli, U. Pliquett, X. Chen, W. Ford, R.J. Swanson, S.J. Beebe, J.F. Kolb, K.H. Schoenbach, Nanosecond pulsed electric fields cause melanomas to self-destruct. Biochem. Biophys. Res. Commun. **343**, 351–360 (2006)
10. S. Su, W. Dai, D.T. Haynie, N. Simicevic, Use of the z-transform to investigate nanopulse penetration of biological matter. Bioelectromagnetics **26**, 389–397 (2005)
11. K. Foster, H. Schwan, Dielectric properties of tissues, in *Biological Effects of Electromagnetic Fields*, ed. By C. Polk, E. Postow (CRC Press, New York, 1996)
12. A. Taflove, S. Hagness, *Computational Electrodynamics: The Finite-Difference Time-Domain Method* (Artech House, Norwood, MA, 2005)
13. D. Caratelli, A. Massaro, R. Cingolani, A. Yarovoy, Accurate time-domain modeling of reconfigurable antenna sensors for non-invasive melanoma skin cancer detection. IEEE Sens. J. **12**, 635–643 (2012)
14. F. Torres, P. Vaudon, B. Jecko, Application of fractional derivatives to the FDTD modelling of pulse propagation in a Cole-Cole dispersive medium. Microw. Opt. Technol. Lett. **13**, 300–304 (1996)
15. Z. Lin, On the FDTD formulations for biological tissues with Cole-Cole dispersion. IEEE Microw. Wirel. Compon. Lett. **20**, 244–246 (2010)
16. H.H. Abdullah, H.A. Elsadek, H.E. ElDeeb, N. Bagherzadeh, Fractional derivatives based scheme for FDTD modeling of th-order Cole-Cole dispersive media. IEEE Antennas Wirel. Propag. Lett. **11**, 281–284 (2012)
17. W.J. Chen, W. Shao, B.Z. Wang, ADE-Laguerre-FDTD method for wave propagation in general dispersive materials. IEEE Microw. Wirel. Compon. Lett. **23**, 228–230 (2013)
18. M.F. Causley, P.G. Petropoulos, On the time-domain response of Havriliak-Negami dielectrics. IEEE Trans. Antennas Propag. **61**, 3182–3189 (2013)
19. S.G. Ha, J. Cho, H. Kim, K.Y. Jung, FDTD dispersive modeling of human tissues based on quadratic complex rational function. IEEE Trans. Antennas Propag. **61**, 996–999 (2013)
20. L. Mescia, P. Bia, D. Caratelli, Authors reply. IEEE Trans. Microw. Theory Tech. **63**, 4191–4193 (2015)
21. D. Caratelli, L. Mescia, P. Bia, O.V. Stukach, Fractional-calculus-based FDTD algorithm for ultra-wideband electromagnetic characterization of arbitrary dispersive dielectric materials. IEEE Trans. Antennas Propag. **64**, 3533–3544 (2016)
22. J. Nelder, R. Mead, A simplex method for function minimization. Comput. J. **7**, 308–313 (1965)
23. J.W. Schuster, R.J.Luebbers, An FDTD algorithm for transient propagation in biological tissue with a Cole-Cole dispersion relation, in *Proceedings of IEEE International Symposium on Antennas and Propagation* (1998), pp. 1988–1991
24. S.D. Gedney, An anisotropic perfectly matched layer absorbing medium for the truncation of FDTD lattices. IEEE Trans. Antennas Propagat. **44**, 1630–1639 (1996)

Spherical Harmonic Solution of the Robin Problem for the Laplace Equation in Supershaped Shells

Diego Caratelli, Pierpaolo Natalini and Paolo Emilio Ricci

Abstract The Robin problem for the Laplace equation in normal-polar shells is addressed by using a suitable spherical harmonic expansion technique. Attention is in particular focused on the wide class of domains whose boundaries are defined by a generalized version of the so-called "superformula" introduced by Gielis. A dedicated numerical procedure based on the computer algebra system Mathematica© is developed in order to validate the proposed methodology. In this way, highly accurate approximations of the solution, featuring properties similar to the classical ones, are obtained.

Keywords Robin problem · Laplace equation · Spherical harmonic expansion · Gielis formula · Supershaped shell

AMS Classification 35J05–35J25

D. Caratelli (✉)
The Antenna Company Nederland B.V., High Tech Campus,
5656 Eindhoven, AE, The Netherlands
e-mail: diego.caratelli@antennacompany.com

D. Caratelli
Tomsk Polytechnic University, 84/3 Sovetskaya Street, 634050 Tomsk, Russia

P. Natalini
Department of Mathematics and Physics, Roma Tre University,
Largo San Leonardo Murialdo, 1, 00146 Rome, Italy
e-mail: natalini@mat.uniroma3.it

P.E. Ricci
International Telematic University UniNettuno, Corso Vittorio
Emanuele II, 39, 00186 Rome, Italy
e-mail: paoloemilioricci@gmail.com

J. Gielis et al. (eds.), *Modeling in Mathematics*,
Atlantis Transactions in Geometry 2, DOI 10.2991/978-94-6239-261-8_2

1 Introduction

Many problems of mathematical physics and electromagnetics are related to the Laplacian differential operator. Among them, it is worth mentioning those relevant to the Laplace and Helmholtz equations. However, most of the mentioned differential problems can be solved in explicit way only in canonical domains with special symmetries, such as intervals, cylinders or spheres [1]. The solution in more general domains can be obtained by using the Riemann theorem on conformal mappings and the relevant invariance of the Laplacian [2]. However, it is clear that conformal mapping techniques can not be used in the three-dimensional case where approaches based on suitable spatial discretization procedures, such as such as finite-difference or finite-element methods, are usually adopted [3].

Different techniques have been proposed in order to solve the mentioned class of differential problems both from a theoretical and numerical point of view (e.g., representing the solution by means of boundary layer techniques [4], solving the corresponding boundary integral equation by iterative methods [5], approximating the relevant Green function by means of the least squares fitting technique [6], solving the linear system relevant to an elliptic partial differential equation by means of relaxation methods [7]). However, none of the contributions already available in the scientific literature deals with the classical Fourier projection method [8] which has been extended in recent papers [9–16] in order to address boundary-value problems (*BVP*s) in simply connected starlike domains whose boundaries may be regarded as an anisotropically stretched unit circle or sphere centered at the origin.

In this contribution, a suitable methodology, based on the theory of spherical harmonics [17], has been developed in order to compute the solution of the Robin problem for the Laplace equation in normal-polar shell-like domains. In particular, the boundaries of the considered domains are supposed to be defined by a generalized version of the so-called Gielis formula (also known as "superformula") [18]. Regular functions are assumed to describe the boundary values, but the proposed approach can be easily generalized in case of weakened hypotheses.

In order to verify and validate the developed technique, a suitable numerical procedure based on the computer algebra system Mathematica© has been adopted. By using such procedure, a point-wise convergence of the spherical harmonic series representation of the solution has been observed in the regular points of the boundaries, with Gibbs-like phenomena potentially occurring in the quasi-cusped points. The obtained numerical results are in good agreement with theoretical findings by Carleson [19].

2 The Laplacian in Stretched Spherical Coordinates

Let us introduce in the real space the usual spherical coordinate system:

$$\begin{cases} x = r \sin\vartheta \cos\varphi\,, \\ y = r \sin\vartheta \sin\varphi\,, \\ z = r \cos\vartheta\,, \end{cases} \tag{1}$$

and the polar equations:

$$r = R^{\pm}(\vartheta, \varphi)\,, \tag{2}$$

relevant to the boundaries of the supershaped shell $\mathcal{S}$ which is described by the following chain of inequalities:

$$R^{-}(\vartheta, \varphi) \le r \le R^{+}(\vartheta, \varphi)\,, \tag{3}$$

with $0 \le \vartheta \le \pi$ and $0 \le \varphi \le 2\pi$. In (2), $R^{\pm}(\vartheta, \varphi)$ are assumed to be piece-wise C^2 functions satisfying the condition:

$$R^{+}(\vartheta, \varphi) > R^{-}(\vartheta, \varphi) > 0\,, \quad 0 \le \vartheta \le \pi\,, \quad 0 \le \varphi \le 2\pi\,. \tag{4}$$

In this way, upon introducing the stretched radius ϱ such that:

$$r = \frac{(b-\varrho)\, R^{-}(\vartheta, \varphi) - (a-\varrho)\, R^{+}(\vartheta, \varphi)}{b-a}\,, \tag{5}$$

with $b > a > 0$, the considered shell-like domain $\mathcal{S}$ can be readily obtained by assuming $0 \le \vartheta \le \pi$, $0 \le \varphi \le 2\pi$, and $a \le \varrho \le b$.

Remark – Note that, in the stretched coordinate system ϱ, ϑ, φ the original domain $\mathcal{S}$ is transformed into the spherical shell of radii a and b, respectively. Hence, in this system one can use classical techniques to solve the Laplace equation, including the eigenfunction method [1].

Let us consider a piece-wise $C^2(\mathring{\mathcal{S}})$ function $v(x, y, z) = v(r \sin\vartheta \cos\varphi, r \sin\vartheta \sin\varphi, r \cos\vartheta) = u(r, \vartheta, \varphi)$ and the Laplace operator, Δ, in spherical coordinates:

$$\Delta u = \frac{1}{r^2}\frac{\partial}{\partial r}\left(r^2 \frac{\partial u}{\partial r}\right) + \frac{1}{r^2 \sin\vartheta}\frac{\partial}{\partial \vartheta}\left(\sin\vartheta \frac{\partial u}{\partial \vartheta}\right) + \frac{1}{r^2 \sin^2\vartheta}\frac{\partial^2 u}{\partial \varphi^2}\,. \tag{6}$$

In the considered stretched coordinate system, Δ can be represented by setting:

$$U(\varrho, \vartheta, \varphi) = u\left(\frac{(b-\varrho)\, R^{-}(\vartheta, \varphi) - (a-\varrho)\, R^{+}(\vartheta, \varphi)}{b-a}, \vartheta\right)\,. \tag{7}$$

In this way, by denoting $R^{\pm}(\vartheta, \varphi)$ as $R^{\pm}$ for the sake of shortness, one can readily find:

$$\frac{\partial u}{\partial r} = \frac{b-a}{R^{+}-R^{-}} \frac{\partial U}{\partial \varrho}, \tag{8}$$

$$\frac{\partial^2 u}{\partial r^2} = \left(\frac{b-a}{R^{+}-R^{-}}\right)^2 \frac{\partial^2 U}{\partial \varrho^2}, \tag{9}$$

$$\frac{\partial u}{\partial \vartheta} = -\frac{R_{\vartheta}}{R^{+}-R^{-}} \frac{\partial U}{\partial \varrho} + \frac{\partial U}{\partial \vartheta}, \tag{10}$$

$$\begin{aligned} \frac{\partial^2 u}{\partial \vartheta^2} &= \frac{2R_{\vartheta}\left(R^{+}_{\vartheta}-R^{-}_{\vartheta}\right) - R_{\vartheta\vartheta}\left(R^{+}-R^{-}\right)}{\left(R^{+}-R^{-}\right)^2} \frac{\partial U}{\partial \varrho} + \\ &+ \left(\frac{R_{\vartheta}}{R^{+}-R^{-}}\right)^2 \frac{\partial^2 U}{\partial \varrho^2} - 2\frac{R_{\vartheta}}{R^{+}-R^{-}} \frac{\partial^2 U}{\partial \varrho \partial \vartheta} + \frac{\partial^2 U}{\partial \vartheta^2}, \end{aligned} \tag{11}$$

$$\begin{aligned} \frac{\partial^2 u}{\partial \varphi^2} &= \frac{2R_{\varphi}\left(R^{+}_{\varphi}-R^{-}_{\varphi}\right) - R_{\varphi\varphi}\left(R^{+}-R^{-}\right)}{\left(R^{+}-R^{-}\right)^2} \frac{\partial U}{\partial \varrho} + \\ &+ \left(\frac{R_{\varphi}}{R^{+}-R^{-}}\right)^2 \frac{\partial^2 U}{\partial \varrho^2} - 2\frac{R_{\varphi}}{R^{+}-R^{-}} \frac{\partial^2 U}{\partial \varrho \partial \varphi} + \frac{\partial^2 U}{\partial \varphi^2}, \end{aligned} \tag{12}$$

with $R = (b-\varrho) R^{-} - (a-\varrho) R^{+}$, and where the subscripts denote the partial differentiation with respect to the polar angles ϑ and φ. Substituting Eqs. (8)–(12) into (6) finally yields:

$$\begin{aligned} \Delta u = \left(\frac{b-a}{R^{+}-R^{-}}\right)^2 &\left\{\left[1+\left(\frac{R_{\vartheta}}{R}\right)^2+\left(\frac{R_{\varphi}}{R\sin\vartheta}\right)^2\right]\frac{\partial^2 U}{\partial \varrho^2} + \frac{R^{+}-R^{-}}{R}\left[2\left(1+\frac{R^{+}_{\vartheta}-R^{-}_{\vartheta}}{R^{+}-R^{-}}\frac{R_{\vartheta}}{R}+\right.\right.\right. \\ &\left.+\frac{R^{+}_{\varphi}-R^{-}_{\varphi}}{R^{+}-R^{-}}\frac{R_{\varphi}}{R\sin^2\vartheta}\right) - \frac{1}{R}\left(R_{\vartheta\vartheta}+R_{\vartheta}\cot\vartheta+\frac{R_{\varphi\varphi}}{\sin^2\vartheta}\right)\left]\frac{\partial U}{\partial \varrho} + \left(\frac{R^{+}-R^{-}}{R}\right)^2\right(\frac{\partial^2 U}{\partial \vartheta^2}+ \\ &\left.\left.+\cot\vartheta\frac{\partial U}{\partial \vartheta}+\frac{1}{\sin^2\vartheta}\frac{\partial^2 U}{\partial \varphi^2}\right) - 2\frac{R^{+}-R^{-}}{R^2}\left(R_{\vartheta}\frac{\partial^2 U}{\partial \varrho \partial \vartheta}+\frac{R_{\varphi}}{\sin^2\vartheta}\frac{\partial^2 U}{\partial \varrho \partial \varphi}\right)\right\}. \end{aligned} \tag{13}$$

As it can be easily noticed, upon setting $R^{-}(\vartheta, \varphi) = a = 0$ and $R^{+}(\vartheta, \varphi) = b = 1$, the classical expression of the Laplacian in spherical coordinates is recovered.

3 The Robin Problem for the Laplace Equation

Let us consider the interior Robin problem for the Laplace equation in a starlike shell $\mathcal{S}$, whose boundaries $\partial^{\pm}\mathcal{S}$ are described by the polar equations $r = R^{\pm}(\vartheta, \varphi)$ respectively:

$$\begin{cases} \Delta v(x, y, z) = 0 , & (x, y, z) \in \mathring{\mathcal{S}} , \\ \lambda^{\pm} v(x, y, z) + \gamma^{\pm} \frac{\partial v}{\partial \nu}(x, y, z) = f^{\pm}(x, y, z) , & (x, y, z) \in \partial^{\pm}\mathcal{S} , \end{cases} \tag{14}$$

where $\hat{\nu}^{\pm} = \hat{\nu}^{\pm}(\vartheta, \varphi)$ denote the outward-pointing normal unit vectors to the domain boundaries $\partial^{\pm}\mathcal{S}$, respectively, and $\lambda^{\pm}$, $\gamma^{\pm}$ are given regular weighting coefficients.

Under the mentioned assumptions, one can prove the following theorem.

Theorem – Let:

$$\psi_{\vartheta}^{\pm}(\vartheta, \varphi) = \frac{R_{\vartheta}^{\pm}(\vartheta, \varphi)}{R^{\pm}(\vartheta, \varphi)} , \tag{15}$$

$$\psi_{\varphi}^{\pm}(\vartheta, \varphi) = \frac{R_{\varphi}^{\pm}(\vartheta, \varphi)}{R^{\pm}(\vartheta, \varphi) \sin \vartheta} , \tag{16}$$

$$\psi^{\pm}(\vartheta, \varphi) = \sqrt{\psi_{\vartheta}^{\pm}(\vartheta, \varphi)^2 + \psi_{\varphi}^{\pm}(\vartheta, \varphi)^2} , \tag{17}$$

and:

$$\begin{aligned} & f^{\pm}\left(R^{\pm}(\vartheta, \varphi) \sin \vartheta \cos \varphi, R^{\pm}(\vartheta, \varphi) \sin \vartheta \sin \varphi, R^{\pm}(\vartheta, \varphi) \cos \vartheta\right) = F^{\pm}(\vartheta, \varphi) = \\ & = \sum_{n=0}^{+\infty} \sum_{m=0}^{n} P_n^m(\cos \vartheta) \left(\alpha_{n,m}^{\pm} \cos m\varphi + \beta_{n,m}^{\pm} \sin m\varphi\right) , \end{aligned} \tag{18}$$

where:

$$\begin{Bmatrix} \alpha_{n,m}^{\pm} \\ \beta_{n,m}^{\pm} \end{Bmatrix} = \epsilon_m \frac{2n+1}{4\pi} \frac{(n-m)!}{(n+m)!} \int_0^{2\pi} \int_0^{\pi} F^{\pm}(\vartheta, \varphi) P_n^m(\cos \vartheta) \begin{Bmatrix} \cos m\varphi \\ \sin m\varphi \end{Bmatrix} \sin \vartheta d\vartheta d\varphi , \tag{19}$$

ϵ_m being the usual Neumann's symbol and $P_n^m(\cdot)$ the associated Legendre function of the first kind and orders n, m. Then, the boundary-value problem (14) for the Laplace equation admits a classical solution $v(x, y, z) \in L^2(\mathcal{S})$ such that the following series expansion holds true:

$$v\left(\frac{R(\varrho,\vartheta,\varphi)}{b-a}\sin\vartheta\cos\varphi,\ \frac{R(\varrho,\vartheta,\varphi)}{b-a}\sin\vartheta\sin\varphi,\ \frac{R(\varrho,\vartheta,\varphi)}{b-a}\cos\vartheta\right)=U(\rho,\vartheta,\varphi)=$$
$$=\sum_{n=0}^{+\infty}\sum_{m=0}^{n}P_n^m(\cos\vartheta)\left[g_{1,n}\left(\frac{R(\varrho,\vartheta,\varphi)}{b-a}\right)(A_{1,n,m}\cos m\varphi+B_{1,n,m}\sin m\varphi)+\right.$$
$$\left.+g_{2,n}\left(\frac{R(\varrho,\vartheta,\varphi)}{b-a}\right)(A_{2,n,m}\cos m\varphi+B_{2,n,m}\sin m\varphi)\right], \tag{20}$$

with:

$$g_{\ell,n}(z)=\begin{cases}z^n, & \ell=1,\\ z^{-n-1}, & \ell=2,\end{cases} \tag{21}$$

For each pair of indices $n\in\mathbb{N}_0:=\mathbb{N}\cup\{0\}$ and $m=0,1,\ldots,n$ define:

$$\begin{bmatrix}\xi_{\ell,n,m}(\vartheta,\varphi)\\ \eta_{\ell,n,m}(\vartheta,\varphi)\end{bmatrix}=g_{\ell,n}\left(R^{\pm}(\vartheta,\varphi)\right)P_n^m(\cos\vartheta)\left\{\lambda^{\pm}\begin{bmatrix}\cos m\varphi\\ \sin m\varphi\end{bmatrix}\pm\frac{\gamma^{\pm}}{\sqrt{1+\psi^{\pm}(\vartheta,\varphi)^2}}\cdot\right.$$
$$\left.\cdot\begin{bmatrix}\cos m\varphi & -\sin m\varphi\\ \sin m\varphi & \cos m\varphi\end{bmatrix}\cdot\begin{bmatrix}\frac{\dot g_{\ell,n}(R^{\pm}(\vartheta,\varphi))}{g_{\ell,n}(R^{\pm}(\vartheta,\varphi))}+\sin\vartheta\frac{\psi_{\vartheta}^{\pm}(\vartheta,\varphi)}{R^{\pm}(\vartheta,\varphi)}\frac{\dot P_n^m(\cos\vartheta)}{P_n^m(\cos\vartheta)}\\ -\frac{m}{\sin\vartheta}\frac{\psi_{\varphi}^{\pm}(\vartheta,\varphi)}{R^{\pm}(\vartheta,\varphi)}\end{bmatrix}\right\}, \tag{22}$$

where $\dot g_{\ell,n}(z)=\frac{d}{dz}g_{\ell,n}(z)$, and:

$$\dot P_n^m(z)=\frac{d}{dz}P_n^m(z)=\frac{nzP_n^m(z)-(n+m)P_{n-1}^m(z)}{z^2-1}. \tag{23}$$

Therefore, the coefficients $A_{\ell,n,m}$, $B_{\ell,n,m}$ in (20) can be determined by solving the infinite linear system:

$$\sum_{n=0}^{+\infty}\sum_{m=0}^{n}\begin{bmatrix}X^-_{1,1,p,q,n,m} & Y^-_{1,1,p,q,n,m} & X^-_{1,2,p,q,n,m} & Y^-_{1,2,p,q,n,m}\\ X^-_{2,1,p,q,n,m} & Y^-_{2,1,p,q,n,m} & X^-_{2,2,p,q,n,m} & Y^-_{2,2,p,q,n,m}\\ X^+_{1,1,p,q,n,m} & Y^+_{1,1,p,q,n,m} & X^+_{1,2,p,q,n,m} & Y^+_{1,2,p,q,n,m}\\ X^+_{2,1,p,q,n,m} & Y^+_{2,1,p,q,n,m} & X^+_{2,2,p,q,n,m} & Y^+_{2,2,p,q,n,m}\end{bmatrix}\cdot\begin{bmatrix}A_{1,n,m}\\ B_{1,n,m}\\ A_{2,n,m}\\ B_{2,n,m}\end{bmatrix}=\begin{bmatrix}\alpha^-_{p,q}\\ \beta^-_{p,q}\\ \alpha^+_{p,q}\\ \beta^+_{p,q}\end{bmatrix}, \tag{24}$$

where:

$$X^{\pm}_{\left\{\substack{1\\2}\right\},\ell,p,q,n,m}=\epsilon_q\frac{2p+1}{4\pi}\frac{(p-q)!}{(p+q)!}\int_0^{2\pi}\int_0^{\pi}\xi^{\pm}_{\ell,n,m}(\vartheta,\varphi)P_p^q(\cos\vartheta)\left\{\begin{matrix}\cos q\varphi\\ \sin q\varphi\end{matrix}\right\}\sin\vartheta d\vartheta d\varphi, \tag{25}$$

$$Y^{\pm}_{\left\{\substack{1\\2}\right\},\ell,p,q,n,m}=\epsilon_q\frac{2p+1}{4\pi}\frac{(p-q)!}{(p+q)!}\int_0^{2\pi}\int_0^{\pi}\eta^{\pm}_{\ell,n,m}(\vartheta,\varphi)P_p^q(\cos\vartheta)\left\{\begin{matrix}\cos q\varphi\\ \sin q\varphi\end{matrix}\right\}\sin\vartheta d\vartheta d\varphi, \tag{26}$$

with $\ell=1,2$ and $p\in\mathbb{N}_0$, $q=0,1,\ldots,p$.

Proof – Upon noting that in the stretched coordinate system $\varrho,\vartheta,\varphi$ introduced in the x,y,z space, the considered domain $\mathcal{S}$ turns into the spherical shell of radii a and

b, one can readily adopt the usual eigenfunction method [1] in combination with the separation of variables (with respect to r, ϑ, and φ). As a consequence, elementary solutions of the problem can be searched in the form:

$$u(r,\vartheta,\varphi) = U\left(\frac{b\left[r - R^{-}(\vartheta,\varphi)\right] - a\left[r - R^{+}(\vartheta,\varphi)\right]}{R^{+}(\vartheta,\varphi) - R^{-}(\vartheta,\varphi)}, \vartheta, \varphi\right) = \mathrm{P}(\rho)\,\Theta(\vartheta)\,\Phi(\varphi)\,. \tag{27}$$

Substituting into the Laplace equation, one easily finds that the functions $\mathrm{P}(\cdot)$, $\Theta(\cdot)$, and $\Phi(\cdot)$ must satisfy the ordinary differential equations:

$$r^2\frac{d^2\mathrm{P}(r)}{dr^2} + 2r\frac{d\mathrm{P}(r)}{dr} - \nu^2\mathrm{P}(r) = 0\,, \tag{28}$$

$$\frac{1}{\sin\vartheta}\frac{d}{d\vartheta}\left[\sin\vartheta\frac{d\Theta(\vartheta)}{d\vartheta}\right] + \left(\nu^2 - \frac{\mu^2}{\sin^2\vartheta}\right)\Theta(\vartheta) = 0\,, \tag{29}$$

$$\frac{d^2\Phi(\varphi)}{d\varphi^2} + \mu^2\Phi(\varphi) = 0\,, \tag{30}$$

respectively. The parameters ν and μ are separation constants, whose choice is governed by the physical requirement that at any fixed point in space the scalar field $u(r,\vartheta,\varphi)$ must be single-valued. So, by setting:

$$\mu = m \in \mathbb{Z}\,, \tag{31}$$

$$\nu^2 = n(n+1) \qquad (n \in \mathbb{N}_0)\,, \tag{32}$$

one can easily find:

$$\Phi(\varphi) = a_m \cos m\varphi + b_m \sin m\varphi\,, \tag{33}$$

$$\Theta(\vartheta) = c_{n,m}P_n^m(\cos\vartheta)\,, \tag{34}$$

$$\mathrm{P}(r) = d_n r^n + e_n r^{-n-1}\,, \tag{35}$$

where $a_m, b_m, c_{n,m}, d_n, e_n \in \mathbb{C}$ denote arbitrary constants. Hence, the general solution of the Robin problem (14) can be searched in the form:

$$u(r,\vartheta,\varphi) = \sum_{n=0}^{+\infty}\sum_{m=0}^{n} P_n^m(\cos\vartheta)\left[r^n\left(A_{1,n,m}\cos m\varphi + B_{1,n,m}\sin m\varphi\right) + \right.$$
$$\left. + r^{-n-1}\left(A_{2,n,m}\cos m\varphi + B_{2,n,m}\sin m\varphi\right)\right]\,. \tag{36}$$

Enforcing the Robin boundary condition yields:

$$F^{\pm}(\vartheta,\varphi) = \lambda^{\pm} u\left(R^{\pm}(\vartheta,\varphi),\vartheta,\varphi\right) + \gamma^{\pm}\frac{\partial u}{\partial \nu}\left(R^{\pm}(\vartheta,\varphi),\vartheta,\varphi\right) = \\ = \lambda^{\pm} u\left(R^{\pm}(\vartheta,\varphi),\vartheta,\varphi\right) + \gamma^{\pm}\nabla u\left(R^{\pm}(\vartheta,\varphi),\vartheta,\varphi\right)\cdot\hat{\nu}^{\pm}(\vartheta,\varphi), \quad (37)$$

where:

$$\nabla u(r,\vartheta,\varphi) = \hat{r}\frac{\partial u(r,\vartheta,\varphi)}{\partial r} + \hat{\vartheta}\frac{1}{r}\frac{\partial u(r,\vartheta,\varphi)}{\partial \vartheta} + \hat{\varphi}\frac{1}{r\sin\vartheta}\frac{\partial u(r,\vartheta,\varphi)}{\partial \varphi}, \quad (38)$$

and:

$$\hat{\nu}^{\pm}(\vartheta,\varphi) = \pm\frac{\hat{r} - \psi^{\pm}_{\vartheta}(\vartheta,\varphi)\,\hat{\vartheta} - \psi^{\pm}_{\varphi}(\vartheta,\varphi)\,\hat{\varphi}}{\sqrt{1+\psi^{\pm}(\vartheta,\varphi)^2}}. \quad (39)$$

Therefore, combining equations above and using a classical harmonic projection method, the Eqs. (22)–(26) follow after some algebraic manipulations. □

It is worth noting that the derived expressions still hold under the assumption that $R^{\pm}(\vartheta,\varphi)$ are piecewise continuous functions, and the boundary values are described by square integrable, not necessarily continuous, functions, so that the relevant spherical harmonic coefficients $\alpha^{\pm}_{n,m}$, $\beta^{\pm}_{n,m}$ in Eq. (19) are finite quantities.

4 Numerical Procedure

In the following numerical examples, let us assume, for the boundaries $\partial^{\pm}\mathcal{S}$ of the considered annulus, general polar equations of the type:

$$R^{\pm}(\vartheta,\varphi) = \left(\left|\frac{1}{d^{\pm}_x}\sin\frac{a^{\pm}_x\vartheta}{2}\cos\frac{b^{\pm}_x\varphi}{4}\right|^{\nu^{\pm}_x} + \left|\frac{1}{d^{\pm}_y}\sin\frac{a^{\pm}_y\vartheta}{2}\sin\frac{b^{\pm}_y\varphi}{4}\right|^{\nu^{\pm}_y} + \left|\frac{1}{d^{\pm}_z}\cos\frac{a^{\pm}_z\vartheta}{2}\right|^{\nu^{\pm}_z}\right)^{-1/\nu^{\pm}_0}, \quad (40)$$

which provides an extension, to the three-dimensional case, of the "superformula" introduced by Gielis in [18]. Very different characteristic geometries, including ellipsoids, ovaloids, and Lamé-type domains (also called "superellipsoids") can be obtained by assuming suitable values of the parameters $a^{\pm}_x$, $a^{\pm}_y$, $a^{\pm}_z$, $b^{\pm}_x$, $b^{\pm}_y$, $d^{\pm}_x$, $d^{\pm}_y$, $d^{\pm}_z$, $\nu^{\pm}_x$, $\nu^{\pm}_y$, $\nu^{\pm}_z$, $\nu^{\pm}_0$ in (40). It is to emphasized that almost all three-dimensional normal-polar shell-like domains can be described, or closely approximated, by the considered formula.

In order to assess the performance of the proposed methodology in terms of numerical accuracy and convergence rate, the relative boundary error has been evaluated as follows:

$$e_N = \sqrt{\frac{\left\| \lambda^- U_N + \gamma^- \frac{\partial U_N}{\partial \nu} - F^- \right\|^2_{\partial^- \mathcal{S}} + \left\| \lambda^+ U_N + \gamma^+ \frac{\partial U_N}{\partial \nu} - F^+ \right\|^2_{\partial^+ \mathcal{S}}}{\| F^- \|^2_{\partial^- \mathcal{S}} + \| F^+ \|^2_{\partial^+ \mathcal{S}}}}, \tag{41}$$

with $\|\cdot\|$ being the usual L^2 norm, and where $U_N = U_N(\varrho, \vartheta, \varphi)$ denotes the partial sum of order N relevant to the spherical harmonic expansion representing the solution of the boundary-value problem for the Laplace equation, namely:

$$U_N(\varrho, \vartheta, \varphi) = \sum_{n=0}^{N} \sum_{m=0}^{n} P_n^m(\cos\vartheta) \left[g_{1,n}\left(\frac{R(\varrho, \vartheta, \varphi)}{b-a} \right) (A_{1,n,m} \cos m\varphi + B_{1,n,m} \sin m\varphi) + \right.$$
$$\left. + g_{2,n}\left(\frac{R(\varrho, \vartheta, \varphi)}{b-a} \right) (A_{2,n,m} \cos m\varphi + B_{2,n,m} \sin m\varphi) \right]. \tag{42}$$

Remark – It is to be noted that, where the boundary values exhibit a rapidly oscillating behavior, the number N of terms in the spherical harmonic expansion approximating the solution of the problem should be increased accordingly in order to achieve the desired numerical accuracy.

4.1 Example 1

Upon assuming in (40) $a_x^{\pm} = a_y^{\pm} = a_z^{\pm} = 2$, $b_x^{\pm} = b_y^{\pm} = 4$, $d_x^- = d_y^- = d_z^- = 4/5$, $d_x^+ = d_y^+ = d_z^+ = 1$, $\nu_x^{\pm} = \nu_y^{\pm} = \nu_z^{\pm} = 3$, and $\nu_0^{\pm} = 2$, the shell $\mathcal{S}_1$ turns to feature a cuboidal shape. Let $f_-(x, y, z) = \sin 2x + \cos 3y + e^{-2y^2-3z^2+1}$ and $f_+(x, y, z) = x^3y^3z^3 + e^{x+y+z} - x + 2y - 3z$ be the functions describing the boundary values. Provided that the weighting coefficients in the Robin condition are $\lambda_- = -2/5$, $\lambda_+ = -1$, $\gamma_- = 1/10$, $\gamma_+ = -1/2$, the relative boundary error e_N as a function of the number N of terms in the truncated series expansion (42) exhibits the behavior shown in Fig. 1. As it appears from Fig. 2, the selection of the expansion order $N = 11$ leads to a very accurate representation ($e_N < 2\%$) of the solution $v_N(x, y, z)$, whose spherical harmonic coefficients $A_{\ell,n,m}$ and $B_{\ell,n,m}$ ($\ell = 1, 2$) are plotted in Fig. 3.

4.2 Example 2

Upon assuming in (40) $a_x^- = a_y^- = a_z^- = b_x^- = b_y^- = 4$, $a_x^+ = a_y^+ = a_z^+ = 2$, $b_x^+ = b_y^+ = 5$, $d_x^- = d_y^- = d_z^- = 1/2$, $d_x^+ = d_y^+ = d_z^+ = 1$, $\nu_x^- = \nu_y^- = \nu_z^{\pm} = \nu_0^{\pm} = 2$, and $\nu_x^+ = \nu_y^+ = 5$, the shell $\mathcal{S}$ turns to feature a star-like shape with inner spherical void. Let $f_-(x, y, z) = \cos(x + y^2 + z^3)$ and $f_+(x, y, z) = \sinh\left(\sqrt[4]{x^2 + y^2 + z^2}\right)$ be the functions describing the boundary values. Provided that the weighting

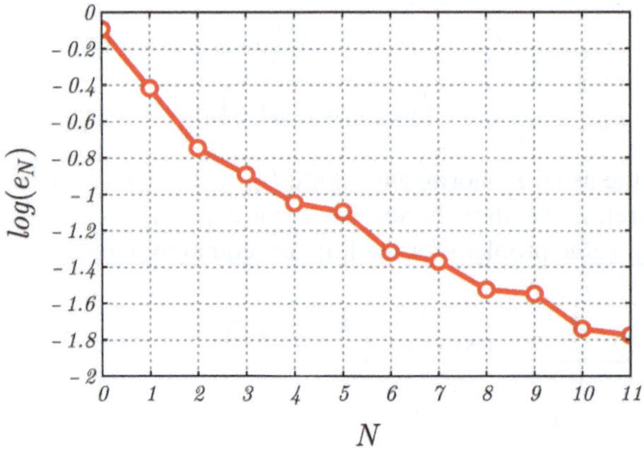

Fig. 1 Relative boundary error e_N as function of the order N of the truncated spherical harmonic expansion representing the solution of the considered Robin problem for the Laplace equation in the supershaped shell $\mathcal{S}_1$ described by the generalized Gielis formula with parameters $a_x^{\pm} = a_y^{\pm} = a_z^{\pm} = 2, b_x^{\pm} = b_y^{\pm} = 4, d_x^- = d_y^- = d_z^- = 4/5, d_x^+ = d_y^+ = d_z^+ = 1, \nu_x^{\pm} = \nu_y^{\pm} = \nu_z^{\pm} = 3, \nu_0^{\pm} = 2$

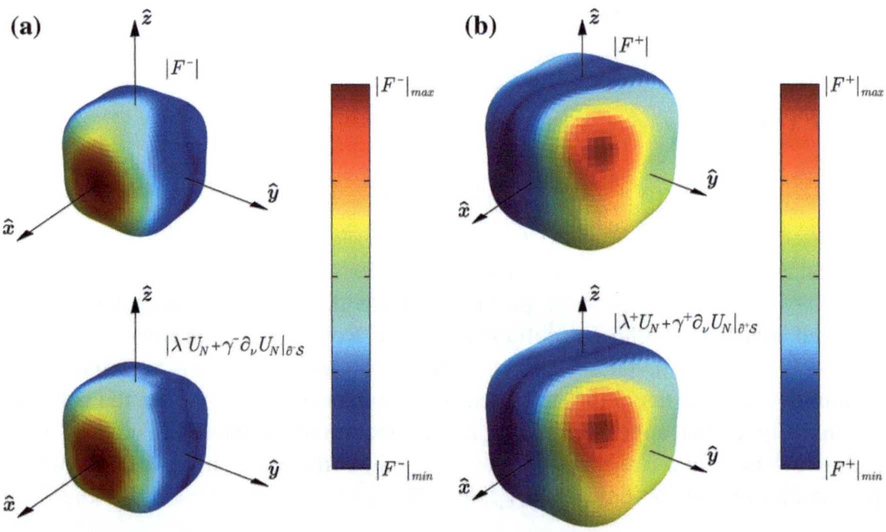

Fig. 2 Boundary behavior along $\partial^- \mathcal{S}$ (**a**) and $\partial^+ \mathcal{S}$ (**b**) of the partial sum U_N $(\varrho, \vartheta, \varphi)$ of order $N = 11$ representing the solution of the considered Robin problem for the Laplace equation in the supershaped shell $\mathcal{S}_1$ described by the generalized Gielis formula with parameters $a_x^{\pm} = a_y^{\pm} = a_z^{\pm} = 2, b_x^{\pm} = b_y^{\pm} = 4, d_x^- = d_y^- = d_z^- = 4/5, d_x^+ = d_y^+ = d_z^+ = 1, \nu_x^{\pm} = \nu_y^{\pm} = \nu_z^{\pm} = 3, \nu_0^{\pm} = 2$

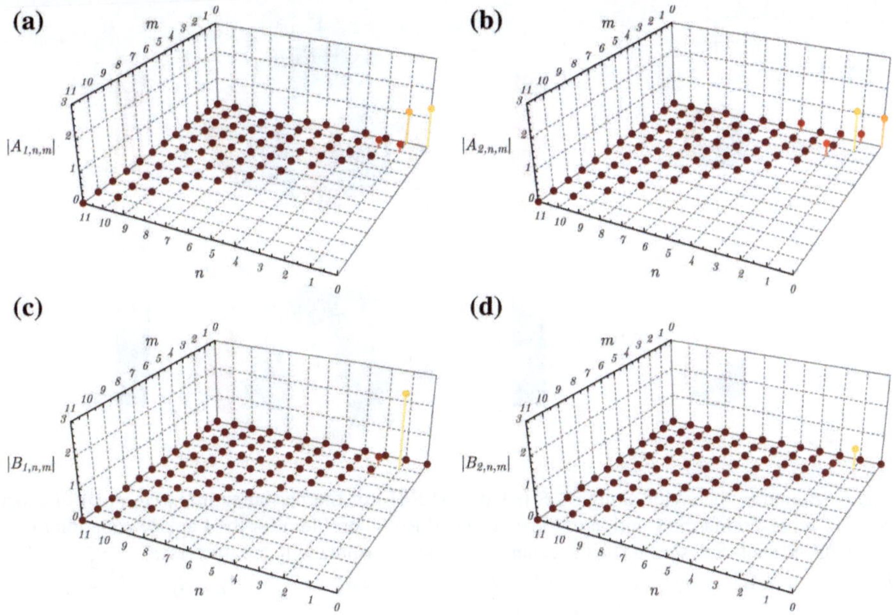

Fig. 3 Magnitude of the coefficients $A_{\ell,n,m}$ and $B_{\ell,n,m}$ ($\ell = 1, 2$) relevant to the spherical harmonic expansion $v_N(x, y, z)$ ($N = 11$)representing the solution of the considered Robin problem for the Laplace equation in the supershaped shell $\mathcal{S}_1$ described by the generalized Gielis formula with parameters $a_x^\pm = a_y^\pm = a_z^\pm = 2$, $b_x^\pm = b_y^\pm = 4$, $d_x^- = d_y^- = d_z^- = 4/5$, $d_x^+ = d_y^+ = d_z^+ = 1$, $\nu_x^\pm = \nu_y^\pm = \nu_z^\pm = 3$, $\nu_0^\pm = 2$

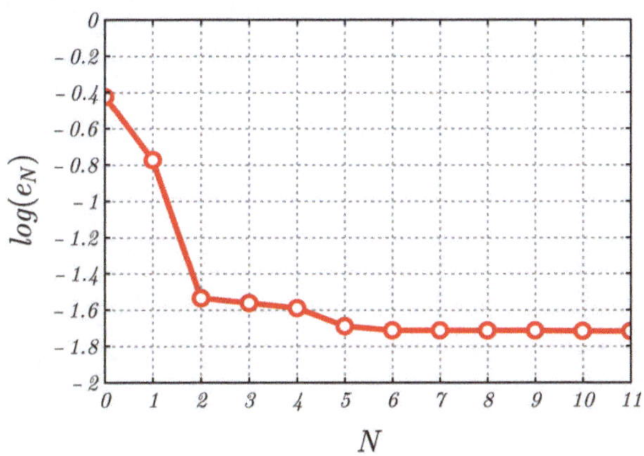

Fig. 4 Relative boundary error e_N as function of the order N of the truncated spherical harmonic expansion representing the solution of the considered Robin problem for the Laplace equation in the supershaped shell $\mathcal{S}$ described by the generalized Gielis formula with parameters $a_x^- = a_y^- = a_z^- = b_x^- = b_y^- = 4$, $a_x^+ = a_y^+ = a_z^+ = 2$, $b_x^+ = b_y^+ = 5$, $d_x^- = d_y^- = d_z^- = 1/2$, $d_x^+ = d_y^+ = d_z^+ = 1$, $\nu_x^- = \nu_y^- = \nu_z^\pm = \nu_0^\pm = 2$, $\nu_x^+ = \nu_y^+ = 5$

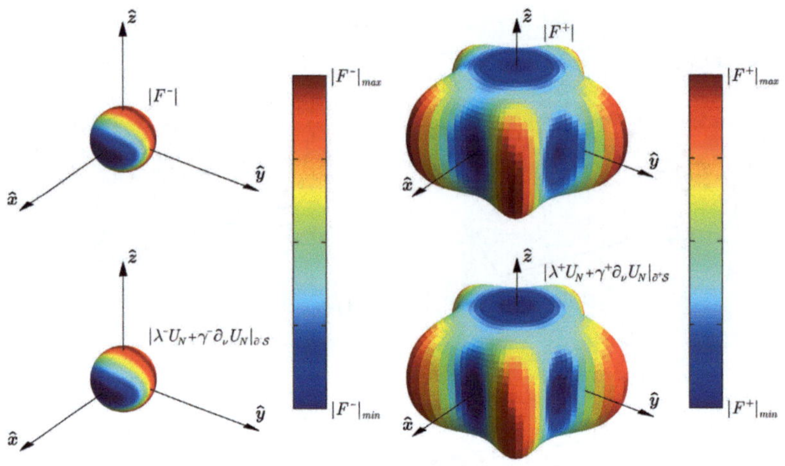

Fig. 5 Boundary behavior along $\partial^- \mathcal{S}$ (**a**) and $\partial^+ \mathcal{S}$ (**b**) of the partial sum $U_N(\varrho, \vartheta, \varphi)$ of order $N = 11$ representing the solution of the considered Robin problem for the Laplace equation in the supershaped shell $\mathcal{S}$ described by the generalized Gielis formula with parameters $a_x^- = a_y^- = a_z^- = b_x^- = b_y^- = 4$, $a_x^+ = a_y^+ = a_z^+ = 2$, $b_x^+ = b_y^+ = 5$, $d_x^- = d_y^- = d_z^- = 1/2$, $d_x^+ = d_y^+ = d_z^+ = 1$, $\nu_x^- = \nu_y^- = \nu_z^\pm = \nu_0^\pm = 2$, $\nu_x^+ = \nu_y^+ = 5$

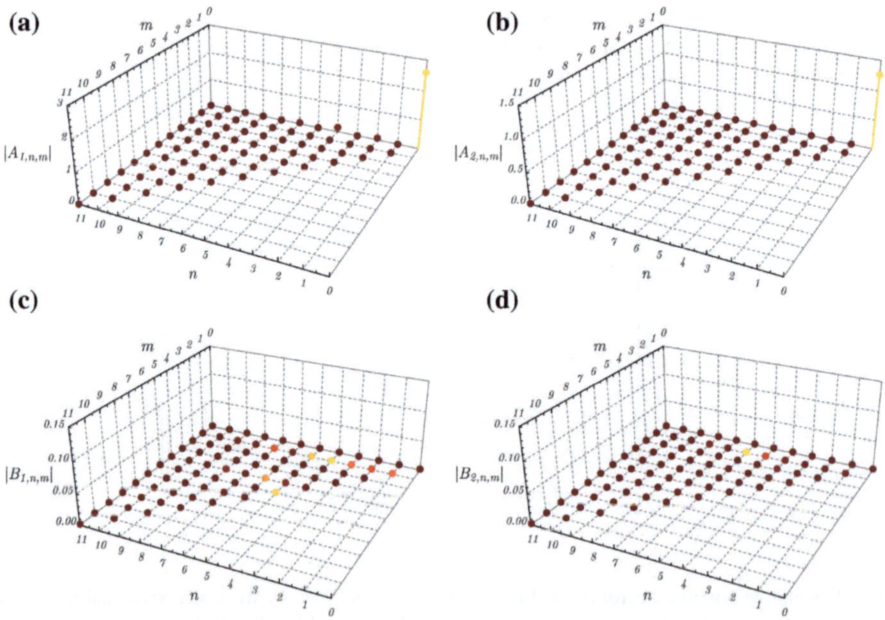

Fig. 6 Magnitude of the coefficients $A_{\ell,n,m}$ and $B_{\ell,n,m}$ ($\ell = 1, 2$) relevant to the spherical harmonic expansion $v_N(x, y, z)$ ($N = 11$)representing the solution of the considered Robin problem for the Laplace equation in the supershaped shell $\mathcal{S}$ described by the generalized Gielis formula with parameters $a_x^- = a_y^- = a_z^- = b_x^- = b_y^- = 4$, $a_x^+ = a_y^+ = a_z^+ = 2$, $b_x^+ = b_y^+ = 5$, $d_x^- = d_y^- = d_z^- = 1/2$, $d_x^+ = d_y^+ = d_z^+ = 1$, $\nu_x^- = \nu_y^- = \nu_z^\pm = \nu_0^\pm = 2$, $\nu_x^+ = \nu_y^+ = 5$

coefficients in the Robin condition are $\lambda_- = -\lambda_+ = 1$, $\gamma_- = -\gamma_+ = -1/5$, the relative boundary error e_N as a function of the number N of terms in the truncated series expansion (42) exhibits the behavior shown in Fig. 4. As it appears from Fig. 5, the selection of the expansion order $N = 11$ leads to a very accurate representation ($e_N < 2\%$) of the solution $v_N(x, y, z)$, whose spherical harmonic coefficients $A_{\ell,n,m}$ and $B_{\ell,n,m}$ ($\ell = 1, 2$) are plotted in Fig. 6.
Remark – It has been observed that L^2 norm of the difference between the exact solution and the relevant approximation is generally negligible. Point-wise convergence seems to be verified in the considered domains, with the only exception of a set of measure zero consisting of quasi-cusped points. In the neighborhood of these points, oscillations of the truncated order solution, recalling the classical Gibbs phenomenon, usually take place.

5 Conclusion

A harmonic projection method, in combination with the adoption of a suitable stretched coordinate system, has been developed for solving the Robin problem for the Laplace equation in supershaped shells. In this way, analytically based expressions of the solution of the considered class of *BVP*s can be derived by using classical quadrature rules, so overcoming the need for cumbersome numerical techniques such as finite-difference or finite-element methods. The proposed approach has been successfully validated by means of a dedicated numerical procedure based on the computer-aided algebra tool Mathematica©. A point-wise convergence of the expansion series representing the solution seems to be verified with the only exception of a set of measure zero consisting of the quasi-cusped points along the boundary of the problem domain. In these points, Gibbs-like oscillations may occur. The computed results are found to be in good agreement with the theoretical findings on Fourier series.

Acknowledgements This study has been partly carried out in the framework of the research and development program running at *The Antenna Company*. For further information, please visit the web site: http://www.antennacompany.com/.

References

1. N.N. Lebedev, *Special Functions and Their Applications* (Dover Inc., New York, 1972)
2. G. Krall, *Meccanica tecnica delle vibrazioni*, vol. II (Veschi, Roma, 1970)
3. A. Bondeson, T. Rylander, P. Ingelstrom, *Computational Electromagnetics* (Springer Science, New York, 2005)
4. D. Medková, Solution of the Dirichlet problem for the Laplace equation. Appl. Math. **44**, 143–168 (1999)
5. B.N. Khoromskiĭ, Integro-difference method of solution of the Dirichlet problem for the Laplace equation. Zh. Vychisl. Mat. i Mat. Fiz. **24**, 53–64 (1984). (in Russian)

6. A. P. Volkov, An effective method for solving the Dirichlet problem for the Laplace equation, (in Russian), *Differentsial'nye Uravneniya*, Vol. 19, 1983, pp. 1000–1007
7. D.M. Young, Iterative methods for solving partial difference equations of elliptic type. Trans. Am. Math. Soc. **76**, 92–111 (1954)
8. G.P. Tolstov, *Fourier Series* (Dover Inc., New York, 1962)
9. P. Natalini, R. Patrizi, P.E. Ricci, The Dirichlet problem for the Laplace equation in a starlike domain of a Riemann surface. Numer. Algorithms **28**, 215–227 (2001)
10. D. Caratelli, P.E. Ricci, The Dirichlet problem for the Laplace equation in a starlike domain, in *Proceedings of International Conference on Scientific Computing*, July 14–17 (Las Vegas, 2008), pp. 160–166
11. D. Caratelli, B. Germano, J. Gielis, M.X. He, P. Natalini, P.E. Ricci, Fourier solution of the Dirichlet problem for the Laplace and Helmholtz equations in starlike domains, in *Lecture Notes of Tbilisi International Centre of Mathematics and Informatics* (Tbilisi University Press, 2010)
12. D. Caratelli, P. Natalini, P.E. Ricci, A. Yarovoy, The Neumann problem for the Helmholtz equation in a starlike planar domain. Appl. Math. Comput. **216**, 556–564 (2010)
13. D. Caratelli, J. Gielis, P. Natalini, P.E. Ricci, I. Tavkelidze, The Robin problem for the Helmholtz equation in a starlike planar domain. Ga. Math. J. **18**, 465–480 (2011)
14. D. Caratelli, J. Gielis, P.E. Ricci, Fourier-like solution of the Dirichlet problem for the Laplace equation in k-type Gielis domains. J. Pure Appl. Math.: Adv. Appl. **5**, 99–111 (2011)
15. D. Caratelli, P.E. Ricci, J. Gielis, The Robin problem for the Laplace equation in a three-dimensional starlike domain. Appl. Math. Comput. **218**, 713–719 (2011)
16. J. Gielis, D. Caratelli, Y. Fougerolle, P.E. Ricci, T. Gerats, Universal natural shapes from unifying shape description to simple methods for shape analysis and boundary value problems, *PLoSOne* (2012) doi:10.1371/journal.pone.0029324
17. G. Dattoli, B. Germano, M.R. Martinelli, P.E. Ricci, A novel theory of Legendre polynomials. Math. Comput. Model. **54**, 80–87 (2011)
18. J. Gielis, A generic geometric transformation that unifies a wide range of natural and abstract shapes. Am. J. Bot. **90**, 333–338 (2003)
19. L. Carleson, On convergence and growth of partial sums of Fourier series. Acta Math. **116**, 135–157 (1966)

Evolution Equations Involving Matrices Raised to Non-Integer Exponents

Giuseppe Dattoli, Silvia Licciardi, Federico Nguyen and Elio Sabia

Abstract The use of matrices evolution equations raised to non-integer exponents finds applications in problems involving the solution of two or three level systems ruled by Klein–Gordon type equations. We develop a fairly simple method exploiting the wealth of results developed on fractional calculus and provide an example of application.

Keywords Fractional calculus · Klein–Gordon equations · Exotic particles · Axion

1 Introduction

In this paper we will discuss the solution of matrix evolution equations, using the formalism of fractional operators, namely of operators raised to a non-integer exponent.

We introduce the topics we will discuss in this paper, its scope and the formalism we will employ by presenting a fairly simple example regarding the solution of the second order differential equation

G. Dattoli (✉) · S. Licciardi · F. Nguyen · E. Sabia
ENEA - Centro Ricerche Frascati, Via Enrico Fermi 45, 00044 Frascati, Rome, Italy
e-mail: giuseppe.dattoli@enea.it

F. Nguyen
e-mail: federico.nguyen@enea.it

E. Sabia
e-mail: elio.sabia@enea.it

S. Licciardi
Department of Mathematics, University of Catania,
Via Santa Sofia 64, 95125 Catania, Italy
e-mail: silvia.licciardi@dmi.unict.it

J. Gielis et al. (eds.), *Modeling in Mathematics*,
Atlantis Transactions in Geometry 2, DOI 10.2991/978-94-6239-261-8_3

$$\frac{d^2}{dt^2}\underline{Y} = -\hat{A}\,\underline{Y}$$
$$\underline{Y}(0) = \underline{Y}_0 \tag{1}$$
$$\frac{d}{dt}\underline{Y}|_{t=0} = \dot{\underline{Y}}_0$$

where $\hat{A}$ is a non-singular 2×2 matrix, with positive defined determinant

$$\hat{A} = \begin{pmatrix} a & b \\ c & d \end{pmatrix} \tag{2}$$

$\underline{Y} = \begin{pmatrix} y_1 \\ y_2 \end{pmatrix}$ a two component column vector and $\underline{Y}_0$, $\dot{\underline{Y}}_0$ the initial conditions of the problem.

The solution of Eq. (1) can be obtained by standard means, namely by introducing a further component $\underline{W} = \frac{d}{dt}\underline{Y}$, thus transforming it into a first order differential equation involving 4×4 matrices. The procedure we will follow foresees different means involving the use of square root matrices. Treating the Eq. (1) as a standard harmonic oscillator equation, we will write the relevant solution as

$$\underline{Y}(t) = e^{it\sqrt{\hat{A}}}\underline{c}_1 + e^{-it\sqrt{\hat{A}}}\underline{c}_2 \tag{3}$$

where $\underline{c}_{1,2}$ are column vectors depending on the initial conditions and will be specified later.

The problem wc are facing with is that of providing an operational meaning to an exponential operator containing the square root of a 2×2 matrix. We proceed therefore as it follows:

(a) Use standard matrix algebra [1] to write

$$\sqrt{\hat{A}} = -\frac{\lambda_-\lambda_+}{\lambda_- - \lambda_+}\left(\frac{1}{\sqrt{\lambda_-}} - \frac{1}{\sqrt{\lambda_+}}\right)\hat{1} + \frac{\sqrt{\lambda_-} - \sqrt{\lambda_+}}{\lambda_- - \lambda_+}\hat{A},$$
$$f_0(\lambda_+, \lambda_-) = -\frac{\lambda_-\lambda_+}{\lambda_- - \lambda_+}\left(\frac{1}{\sqrt{\lambda_-}} - \frac{1}{\sqrt{\lambda_+}}\right), \tag{4}$$
$$f_1(\lambda_+, \lambda_-) = \frac{\sqrt{\lambda_-} - \sqrt{\lambda_+}}{\lambda_- - \lambda_+}$$

with $\hat{1}$, $\lambda_\pm$ being the unit matrix and the eigenvalues of the matrix $\hat{A}$.

(b) Write then

$$e^{\tau\sqrt{\hat{A}}} = e^{f_0(\lambda_+,\lambda_-)\tau\,\hat{1}}e^{f_1(\lambda_+,\lambda_-)\,\tau\,\hat{A}}, \tag{5}$$

(c) Use the Cayley–Hamilton theorem to write the explicit form of the matrix exponential as (see refs. [1, 2])

$$
\begin{aligned}
e^{\tau\sqrt{\hat{A}}} &= e^{f_0(\lambda_+,\lambda_-)\tau}\begin{pmatrix} 1 & 0 \\ 0 & 1 \end{pmatrix}\begin{pmatrix} e_{1,1} & e_{1,2} \\ e_{2,1} & e_{2,2} \end{pmatrix},\\
e_{1,1} &= \left[\frac{(a-d)\,\tau}{\sqrt{\Delta}} f_1(\lambda_+,\lambda_-)\sinh\left(\frac{\sqrt{\Delta}}{2}\right) + \cosh\left(\frac{\sqrt{\Delta}}{2}\right)\right] e^{\frac{1}{2}f_1(\lambda_+,\lambda_-)\,(a+d)\tau}\\
e_{2,2} &= \left[-\frac{(a-d)\,\tau}{\sqrt{\Delta}} f_1(\lambda_+,\lambda_-)\sinh\left(\frac{\sqrt{\Delta}}{2}\right) + \cosh\left(\frac{\sqrt{\Delta}}{2}\right)\right] e^{\frac{1}{2}f_1(\lambda_+,\lambda_-)\,(a+d)\tau},\\
e_{1,2} &= f_1(\lambda_+,\lambda_-)\frac{2\,b\tau}{\sqrt{\Delta}}\sinh\left(\frac{\sqrt{\Delta}}{2}\right) e^{\frac{1}{2}f_1(\lambda_+,\lambda_-)\,(a+d)\tau},\\
e_{2,1} &= f_1(\lambda_+,\lambda_-)\frac{2\,c\tau}{\sqrt{\Delta}}\sinh\left(\frac{\sqrt{\Delta}}{2}\right) e^{\frac{1}{2}f_1(\lambda_+,\lambda_-)\,(a+d)\tau},\\
\Delta &= (f_1(\lambda_+,\lambda_-)\,\tau)^2\left((a-d)^2+4bc\right)
\end{aligned}
\tag{6}
$$

The solution of Eq. (1) can accordingly be written as

$$
\begin{aligned}
\underline{Y}(t) &= C\left(t\sqrt{\hat{A}}\right)\underline{Y}_0 + \frac{1}{\sqrt{\hat{A}}}S\left(t\sqrt{\hat{A}}\right)\underline{\dot{Y}}_0,\\
C\left(t\sqrt{\hat{A}}\right) &= \frac{e^{i\,t\sqrt{\hat{A}}} + e^{-i\,t\sqrt{\hat{A}}}}{2},\\
S\left(t\sqrt{\hat{A}}\right) &= \frac{e^{i\,t\sqrt{\hat{A}}} - e^{-i\,t\sqrt{\hat{A}}}}{2i}
\end{aligned}
\tag{7}
$$

where $C\left(t\sqrt{\hat{A}}\right)$, $S\left(t\sqrt{\hat{A}}\right)$ are pseudo oscillating cos and sin- like solutions.

A graphical example of such solution is shown in Fig. 1, where we have reported the time behavior of the two components $y_{1,2}$ for different values of the matrix entries: (a) $\hat{A}_1 = \begin{pmatrix} 1 & 0.5 \\ 0.5 & 1 \end{pmatrix}$; (b) $\hat{A}_2 = \begin{pmatrix} 1.7 & 0.5 \\ 0.5 & 1.8 \end{pmatrix}$.

For completeness sake we have reported in Fig. 2 the so-called Lissajous curves arising from plotting $S\left(t\sqrt{\hat{A}}\right)$ vs $C\left(t\sqrt{\hat{A}}\right)$.

We will discuss the application of the method to a physical problem later in the paper and concentrate on further refinement of the relevant mathematical details. The paper consists of four sections, in Sect. 2 we will extend the method including the integral transform technique and the square root of to higher order matrices, in Sect. 3 we will discuss a relevant application and Sect. 4 is devoted to final comments and to further possible developments.

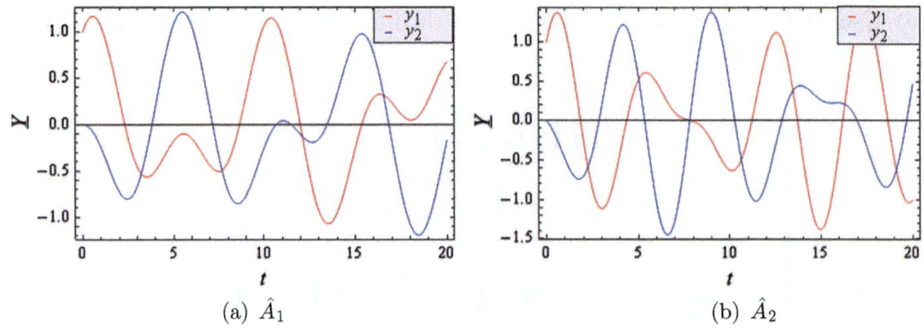

Fig. 1 $\underline{Y}$ versus time; in *red* and *blue* the components y_1, y_2 respectively, with initial conditions $\underline{Y}_0 = \binom{1}{0}$ and $\underline{\dot{Y}}_0 = \binom{-0.1}{0.2}$

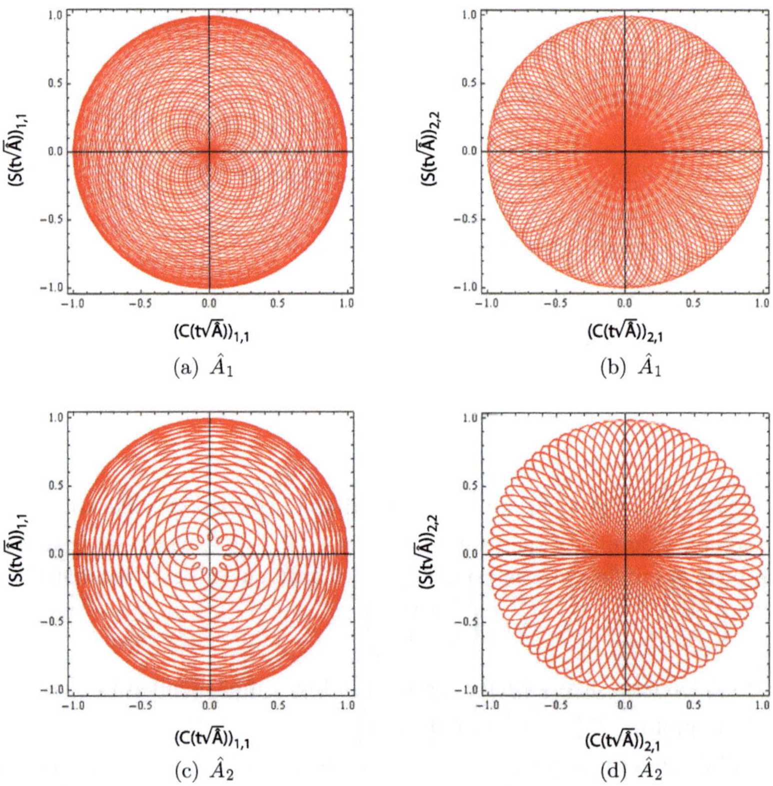

Fig. 2 $S\left(t\sqrt{\hat{A}}\right)$ versus $C\left(t\sqrt{\hat{A}}\right)$, with different components combinations for $\hat{A}_1$ (**a, b**) and $\hat{A}_2$ (**c, d**)

2 Fractional Matrix Exponentiation

The results so far obtained will be further elaborated, including the extension to higher dimensions, but, before addressing this specific aspect of the discussion, we will complement the previous treatment by using techniques developed within the context of fractional calculus [3] and more specifically with the context of evolution problems regarding the solution of evolution equations, like Bethe–Salpeter or other relativistic forms, involving square roots of differential operators [4].

The theory of differ-integral calculus, namely of derivatives and integrals of non-integer order, has received a significant support from methods associated with the Laplace transform and, within such a framework, the use of Lévy transform [5] has been proved to be an effective tool to deal with exponentials with arguments consisting of fractional operators. To this aim we remind that

$$\begin{aligned} e^{-p^{\frac{1}{2}}} &= \int_0^{\infty} g_{\frac{1}{2}}(\eta)\, e^{-\eta\, p} d\eta, \\ g_{\frac{1}{2}}(\eta) &= \frac{1}{2\sqrt{\pi\eta^3}} e^{-\frac{1}{4\eta}} \end{aligned} \tag{8}$$

with $g_{\frac{1}{2}}(\eta)$ being the Lévy-Smirnov distribution.
According to Eq. (8) we can express the exponential of a square root matrix in terms of the integral transform

$$e^{-\tau\sqrt{\hat{A}}} = \int_0^{\infty} g_{\frac{1}{2}}(\eta)\, e^{-\eta\tau^2 \hat{A}} d\eta \tag{9}$$

The advantage offered by Eq. (9) is that the exponential inside the integral (9) depends on the matrix $\hat{A}$ without any further exponentiation, this allows a direct use of the Cayley–Hamilton theorem, to solve the problem of getting an explicit expression for the l.h.s. of Eq. (9). Without considering the general case, reported in Sect. 4, we discuss a few interesting examples

$$\hat{A} = \hat{i} = \begin{pmatrix} 0 & -1 \\ 1 & 0 \end{pmatrix} \tag{10}$$

namely the "unit circular" matrix satisfying the identity

$$\hat{i}^2 = -\hat{1} \tag{11}$$

and generalizing the Euler identity

$$e^{\hat{i}\vartheta} = \cos(\vartheta)\,\hat{1} + \sin(\vartheta)\,\hat{i} \tag{12}$$

The use of Eq. (6) yields

$$e^{-\tau\sqrt{\hat{i}}} = e^{-\frac{\sqrt{2}}{2}\tau}\begin{pmatrix}\cos\left(\frac{\sqrt{2}}{2}\tau\right) & -\sin\left(\frac{\sqrt{2}}{2}\tau\right)\\ \sin\left(\frac{\sqrt{2}}{2}\tau\right) & \cos\left(\frac{\sqrt{2}}{2}\tau\right)\end{pmatrix} \tag{13}$$

representing a kind of damped matrix rotation.

Using an almost similar argument we find that the square root of the "unit hyperbolic matrix"

$$\hat{h} = \begin{pmatrix}0 & 1\\ 1 & 0\end{pmatrix} \tag{14}$$

can be written as

$$\sqrt{\hat{h}} = \frac{1}{2}\left[(1+i)\,\hat{1} + (1-i)\,\hat{h}\right] \tag{15}$$

The solution of equations like

$$\frac{d^2}{d\,\tau^2}\underline{Y} = \hat{h}\,\underline{Y} \tag{16}$$

can accordingly be written as

$$\begin{aligned}\underline{Y}(\tau) &= \begin{pmatrix}\cosh(\tau_+)\cosh(\tau_-) & \sinh(\tau_+)\sinh(\tau_-)\\ \sinh(\tau_+)\sinh(\tau_-) & \cosh(\tau_+)\cosh(\tau_-)\end{pmatrix}\underline{Y}_0 + \\ &+ \frac{1}{\sqrt{\hat{h}}}\begin{pmatrix}\sinh(\tau_+)\cosh(\tau_-) & \cosh(\tau_+)\sinh(\tau_-)\\ \cosh(\tau_+)\sinh(\tau_-) & \sinh(\tau_+)\cosh(\tau_-)\end{pmatrix}\dot{\underline{Y}}_0, \\ \tau_\pm &= \frac{1}{2}(1 \pm i)\,\tau\end{aligned} \tag{17}$$

It is evident that the technique we have envisaged can be extended to higher order matrices, the only problem is that the procedure becomes slightly more cumbersome from the analytical point of view, but it is easily implemented with Mathematica™.

In the case in which $\hat{A}$ is a 3×3 matrix we find

$$e^{\tau\sqrt{\hat{A}}} = e^{f_0(\lambda_1,\lambda_2,\lambda_3)\tau\,\hat{1}}\,e^{f_1(\lambda_1,\lambda_2,\lambda_3)\,\tau\,\hat{A}}\,e^{f_2(\lambda_1,\lambda_2,\lambda_3)\,\tau\,\hat{A}^2} \tag{18}$$

with $\lambda_{1,2,3}$ being the associated eigenvalues and

$$\begin{pmatrix}f_0\\ f_1\\ f_2\end{pmatrix} = \begin{pmatrix}1 & \lambda_1 & \lambda_1^2\\ 1 & \lambda_2 & \lambda_2^2\\ 1 & \lambda_3 & \lambda_3^2\end{pmatrix}^{-1}\begin{pmatrix}\sqrt{\lambda_1}\\ \sqrt{\lambda_2}\\ \sqrt{\lambda_3}\end{pmatrix} \tag{19}$$

The explicit form of the matrix can be written in terms of the exponential of the matrix $\hat{A}$ if we use the identity

$$e^{f_2\,\tau\hat{A}^2} = \frac{1}{\sqrt{\pi}}\int_{-\infty}^{+\infty} e^{-\xi^2+2\xi\sqrt{f_2\tau}\hat{A}}d\xi \tag{20}$$

which yields

$$e^{\tau\sqrt{\hat{A}}} = \frac{e^{\tau\,f_0}}{\sqrt{\pi}}\begin{pmatrix}1 & 0 & 0\\ 0 & 1 & 0\\ 0 & 0 & 1\end{pmatrix}\int_{-\infty}^{+\infty} e^{-\xi^2+(f_1\tau+2\xi\sqrt{f_2\tau})\hat{A}}d\xi \tag{21}$$

A fairly simple example showing the effectiveness of the method is provided by the matrix

$$\hat{A} = \begin{pmatrix}0 & -\omega_3 & \omega_2\\ \omega_3 & 0 & -\omega_1\\ -\omega_2 & \omega_1 & 0\end{pmatrix} \tag{22}$$

whose exponentiation yields a Rodrigues matrix $\hat{R}$ [1, 6]

$$\begin{aligned} e^{\vartheta\,\hat{A}} &= \hat{R}(\vartheta) = \hat{1} + \vartheta\,\mathrm{sinc}\,(\Omega\,\vartheta)\,\hat{A} + \frac{1}{2}\vartheta^2\left[\mathrm{sinc}\left(\frac{\Omega\,\vartheta}{2}\right)\right]^2\hat{A}^2,\\ \Omega &= \sqrt{\omega_1^2+\omega_2^2+\omega_3^2} \end{aligned} \tag{23}$$

The use of the previous identity and a Gaussian integration, finally yields the following factorization

$$e^{\tau\sqrt{\hat{A}}} = e^{\tau\,f_0}\left[\hat{1} + \left(\frac{1-e^{-\Omega^2 f_2\tau}}{\Omega^2}\right)\hat{A}^2\right]\hat{R}(\tau\,f_1) \tag{24}$$

The extension to higher order matrices will be discussed in the final remarks, containing further comments on the technique we have proposed.

3 A Physical Application

As previously noted, the use of pseudo differential operators for the solution of Klein–Gordon or Bethe–Salpeter equations has gained considerable interest in the last few years, we will use the methods discussed in the previous sections to study the solution of the stationary Klein–Gordon equation describing the axion photon coupling in a transverse magnetic field. According to the analysis developed in ref. [7] photon γ and axion a fields can be viewed as two state polarization, which, propagating in an intense magnetic field, undergo a kind of Cotton-Mouton rotation.

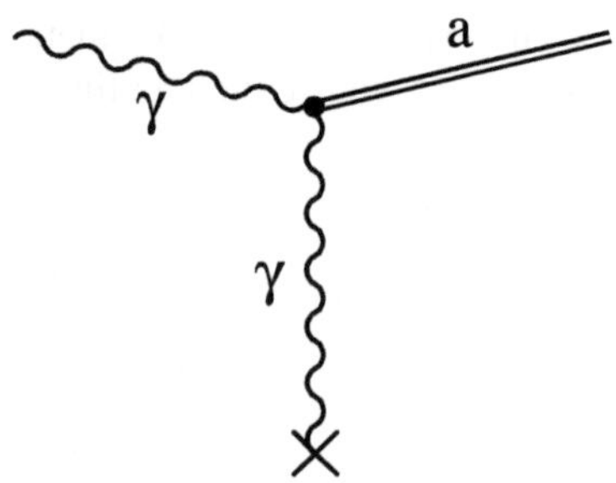

Fig. 3 Primakoff process for photon (γ) - photon (γ) - axion (a) production

Within such a framework the photon-axion interaction can be viewed as a kind of Primakoff process (see Fig. 3) in which the vertex coupling occur between a real external photon, the virtual photon associated with the static magnetic field and the axion.

The equation we will consider is [7]

$$\left(\frac{\partial}{\partial t}\right)^2 \begin{pmatrix}\gamma\\ a\end{pmatrix} = -\omega^2 \begin{pmatrix} 2n-1 & \frac{g_{a\gamma}B}{\omega} \\ \frac{g_{a\gamma}B}{\omega} & 1-\frac{m_a^2}{\omega^2}\end{pmatrix}\begin{pmatrix}\gamma\\ a\end{pmatrix} \tag{25}$$

where n is the refractive photon index associated to the magnetic field B, $g_{a\gamma}$ the axion photon coupling constant, m_a is the axion mass and ω is the external photon frequency. Without further considering the specific details, we find that the eigenvalues of the matrix are

$$\begin{aligned} \lambda_\pm &= \frac{N+M\pm\sqrt{[N-M]^2+4G^2}}{2}, \\ N &= 2n-1,\ M = 1-\left(\frac{m_a}{\omega}\right)^2,\ G = \frac{g_{a\gamma}B}{\omega} \end{aligned} \tag{26}$$

If we make the assumption that $\frac{m_a}{\omega} << 1$ (which means that the energy of the external photon is much larger than the axion mass [8]) and $N \cong 1$ we are left with $\lambda_\pm = 1 \pm G$ furthermore by noting also that $G << 1$, we find

$$\begin{aligned} f_0(\lambda_+,\lambda_-) &= \frac{1-G^2}{2G}\left(\frac{1}{\sqrt{1-G}} - \frac{1}{\sqrt{1+G}}\right) \cong \frac{1}{2} - \frac{3}{16}G^2 \\ f_1(\lambda_+,\lambda_-) &= \frac{\sqrt{1-G}-\sqrt{1+G}}{-2G} \cong \frac{1}{2} + \frac{G^2}{16} \end{aligned} \tag{27}$$

and

$$e^{i\omega t\sqrt{\hat{A}}} = e^{-i\omega t[(1-\frac{1}{8}G^2)]}\begin{pmatrix}\cos(f_1 G\omega t) & -\sin(f_1 G\omega t)\\ \sin(f_1 G\omega t) & \cos(f_1 G\omega t)\end{pmatrix} \tag{28}$$

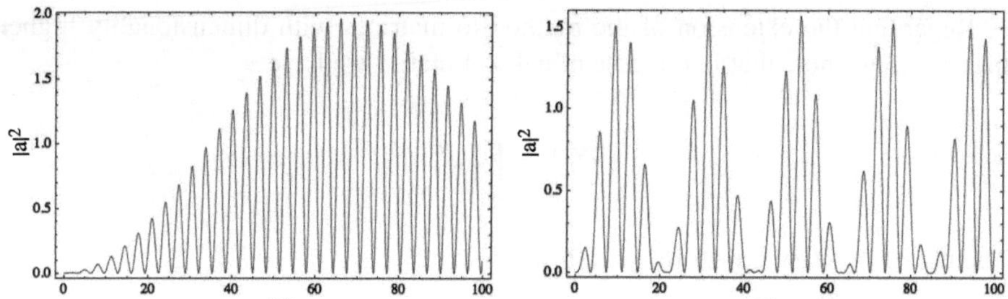

(a) Case 1 (b) Case 2

Fig. 4 Axion generation probability ($| a |^2 \cdot 10^{15}$) versus ωt for different parameters

which is essentially a rotation matrix induced by the axion-photon coupling constant. The validity of our solution is limited to the case in which both $\lambda_\pm$ are non-negative, therefore the following conditions should be satisfied

$$m_a \leq \omega \sqrt{1 - \frac{1}{2n-1}\left(\frac{g_{a\gamma}B}{\omega}\right)^2} \tag{29}$$

It is interesting to look at the behavior vs. time of the probability of creating an axion during the interaction for different values of the parameter. In Fig. 4 we have shown $|a|^2$ vs $\omega\, t$ for the cases

(1) $\frac{g_{\alpha\gamma}B}{\omega} \cong 2 \cdot 10^{-9}, \quad \left(\frac{m_a}{\omega}\right) \cong 0.3;$
(2) $\frac{g_{\alpha\gamma}B}{\omega} \cong 10^{-8}, \quad \left(\frac{m_a}{\omega}\right) \cong 0.71.$

The probability of converting an axion into a photon displays an analogous behaviour vs. time. The combined processes of photon-axion-photon conversion are the pivoting tools of shining through the wall experiments [7, 8], which are of noticeable importance for the search of dark matter constituents.

Further details on the use of the above formalism for the study of a facility dedicated to axion research can be found elsewhere [9].

4 Final Comments

In these concluding remarks we want to emphasize that the mathematical tool we have developed is amenable for further extensions and contains seed elements for further studies.

Regarding the extension of the method to matrices with dimensionality higher than 3 × 3 we note that in the case of a 4 × 4 matrix we have

$$e^{\tau\sqrt{\hat{A}}} = \prod_{r=0}^{3} e^{f_r \tau \hat{A}^r} \tag{30}$$

The use of the Airy transform [2] allows to write

$$\begin{aligned} e^{\alpha \hat{A}^3} &= \int_{-\infty}^{\infty} Ai(t)\, e^{\sqrt[3]{\alpha} t \hat{A}} dt, \\ Ai(t) &= \frac{1}{2\pi} \int_{-\infty}^{\infty} e^{\left(\frac{i}{3}\xi^3 + it\xi\right)} d\xi \end{aligned} \tag{31}$$

Thus finally getting, for the square root of a 4 × 4 matrix, the following expression in terms of Airy and Gauss transforms:

$$\begin{aligned} e^{\tau\sqrt{\hat{A}}} &= \frac{e^{f_0(\tau)}}{\sqrt{\pi}} \int_{-\infty}^{+\infty} \int_{0}^{\infty} e^{-\xi^2} Ai(\lambda) \hat{U}(f_1\tau + 2\sqrt{f_2\tau}\xi + \sqrt[3]{f_3\tau}\lambda) d\xi d\lambda \\ \hat{U}(\tau) &= e^{\tau \hat{A}} \end{aligned} \tag{32}$$

The case of matrices with higher order dimensionalities (5 × 5, ...), requires only higher order integral transforms of the type discussed in ref. [10].

The use of the Lévy-Smirnov distribution allows to write (see Eq. (23))

$$e^{-\tau\sqrt{\begin{pmatrix} 0 & -\omega_3 & \omega_2 \\ \omega_3 & 0 & -\omega_1 \\ -\omega_2 & \omega_1 & 0 \end{pmatrix}}} = \int_0^{\infty} g_{\frac{1}{2}}(\eta)\, \hat{R}(-\eta\,\tau^2) d\eta, \tag{33}$$

Taking into account that

$$\begin{aligned} &\int_0^{\infty} g_{\frac{1}{2}}(\eta)\, d\eta = 1, \\ &\int_0^{\infty} g_{\frac{1}{2}}(\eta) e^{-i\,\eta x}\, d\eta = e^{-\frac{\sqrt{2x}}{2}} \left(\cos\left(\frac{\sqrt{2x}}{2}\right) - i \sin\left(\frac{\sqrt{2x}}{2}\right)\right) \end{aligned} \tag{34}$$

and, by using the explicit form of the Rodrigues matrix (see Eqs. (22) and (23)) we end up with

$$\begin{aligned} &e^{-\tau\sqrt{\hat{A}}} = \\ &= \hat{1} - \frac{e^{-\frac{\sqrt{2\Omega}}{2}\tau}}{\Omega} \sin\left(\frac{\sqrt{2\Omega}}{2}\tau\right) \hat{A} + \frac{1}{\Omega^2}\left(1 - e^{-\frac{\sqrt{2\Omega}}{2}\tau} \cos\left(\frac{\sqrt{2\Omega}}{2}\tau\right)\right) \hat{A}^2 \end{aligned} \tag{35}$$

The possibility of extending the method to higher order roots follows the same procedure we have envisaged and for example

$$e^{-\tau \sqrt[n]{\hat{A}}} = \int_0^\infty g_{\frac{1}{n}}(\eta)\, e^{-\eta \tau^n \hat{A}} d\eta \tag{36}$$

where the Lévy stable function $g_{\frac{1}{n}}(\eta)$ is in general expressible in terms of Meijer G-functions (see ref. [5]). A more detailed account of this aspect of the problem will be given elsewhere.

References

1. G. Birkhoff, S. Mac Lane, *A Survey of Modern Algebra*, 3rd edn. (Macmillan, New York, 1965). Chapter 6
2. D. Babusci, G. Dattoli, M. Del Franco, *Lectures on Mathematical Methods for Physics*. RT/2010/58/ENEA
3. R. Herman, *Fractional Calculus: An Introduction for Physicist* (World Scientific, Singapore, 2011)
4. D. Babusci, G. Dattoli, M. Quattromini, Phys. Rev. A **83**, 062109 (2011); Kowalski, J. Rembielinski. Phys. Rev. A **84**, 012108 (2011)
5. K. Penson, K. Gorska, Phys. Rev. Lett. **105**, 210604 (2010)
6. R.M. Murray, Z. Li, S.S. Sastry, *A Mathematical Introduction to Robotic Manipulation* (CRC Press, Boca Raton, 1994)
7. G. Raffelt, L. Stodolsky, Phys. Rev. D **37**, 1237 (1988)
8. K.A. Olive et al., Particle Data Group. Chin. Phys. C **38**, 090001 (2014). and 2015 update
9. G. Dattoli et al., *Undulator and Free Electron Laser radiation for Fundamental Physics research*, talk at the 7th International Conference Charged and Neutral Particles Channeling Phenomena Channeling 2016, https://agenda.infn.it/conferenceOtherViews.py?view=standard&confId=10663
10. D. Babusci, G. Dattoli, D. Sacchetti, The airy transform and the associated polynomials. Centr. Eur. J. Phys. **9**, 1381 (2011). doi:10.2478/s11534-011-0057-9

Hermite Calculus

Giuseppe Dattoli, Bruna Germano, Silvia Licciardi and Maria Renata Martinelli

Abstract We develop a new method of umbral nature to treat blocks of Hermite and of Hermite like polynomials as independent algebraic quantities. The Calculus we propose allows the formulation of a number of "practical rules" yielding significant simplifications in computational problems involving integrals and partial differential equations as well. The procedure we adopt is particularly useful to enter more deeply in the algebraic structure of Hermite polynomials. It provides indeed a tool allowing a generalization of the recently introduced geometrical point of view to the interplay between ordinary monomials and Hermite polynomials.

1 Introduction

In this paper we deal with a protocol, which will be referred as Hermite calculus, useful to treat computations involving Hermite polynomials and their generalizations as well.

G. Dattoli (✉) · S. Licciardi
ENEA - Frascati Research Center, Via Enrico Fermi 45, 00044 Frascati, Rome, Italy
e-mail: giuseppe.dattoli@enea.it

B. Germano · M.R. Martinelli
Department of Methods and Mathematic Models for Applied Sciences,
University of Rome, La Sapienza, Via A. Scarpa, 14, 00161 Rome, Italy
e-mail: bruna.germano@sbai.uniroma1.it

M.R. Martinelli
e-mail: martinelli@dmmm.uniroma1.it

S. Licciardi
Department of Mathematics, University of Catania, Via Santa Sofia,
64, 95125 Catania, Italy
e-mail: silvia.licciardi@dmi.unict.it

J. Gielis et al. (eds.), *Modeling in Mathematics*,
Atlantis Transactions in Geometry 2, DOI 10.2991/978-94-6239-261-8_4

To give a flavour of the techniques we will employ we consider the integral

$$I(\alpha,\beta,\gamma)=\int_{-\infty}^{\infty}e^{-(\alpha+\beta)x^2-\gamma x}dx \tag{1}$$

which can be evaluated with ordinary means, thus getting

$$I(\alpha,\beta,\gamma)=\sqrt{\frac{\pi}{\alpha+\beta}}e^{\frac{\gamma^2}{4(\alpha+\beta)}} \tag{2}$$

We will test the formalism, we are going to decribe, by using such a benchmark and restyle Eq. (1) as it follows

$$I(\alpha,\beta,\gamma)=\int_{-\infty}^{\infty}e^{-\alpha x^2-\hat{h}_{(\gamma,-\beta)}x}dx \tag{3}$$

where we have introduced the notation

$$e^{-\hat{h}_{(\gamma,-\beta)}x}=\sum_{r=0}^{\infty}\frac{(-x)^r}{r!}\hat{h}^r_{(\gamma,-\beta)}=\sum_{r=0}^{\infty}\frac{(-x)^r}{r!}H_r(\gamma,-\beta) \tag{4}$$

based on the use of the umbral identity [1, 2][1]

$$\begin{aligned}&\hat{h}^r_{(\gamma,-\beta)}=H_r(\gamma,-\beta),\\ &H_r(x,y)=r!\sum_{s=0}^{[\frac{r}{2}]}\frac{x^{r-2s}y^s}{(r-2s)!s!}\end{aligned} \tag{5}$$

In the integral in Eq. (4) we have treated the term which can be expended in terms of Hermite polynomials as a single block and we have enucleated the variable x raised to the first power.

We will now follow a prescription according to which the operator $\hat{h}$ is treated as an ordinary algebraic quantity.

According to the ordinary rules for the Gaussian integrals we can write [2, 3]

$$I(\alpha,\beta,\gamma)=\sqrt{\frac{\pi}{\alpha}}e^{\frac{\hat{h}^2_{(\gamma,-\beta)}}{4\alpha}}=\sqrt{\frac{\pi}{\alpha}}\sum_{r=0}^{\infty}\frac{1}{r!}\left(\frac{\hat{h}^2_{(\gamma,-\beta)}}{4\alpha}\right)^r \tag{6}$$

which provides us with the correct result for the problem we are studying. The application of the previous prescription yields, indeed, if $\left|\frac{\beta}{\alpha}\right|<1$

[1] We have used the umbral notation in a rather inaccurate way, without specifying on which space the operators are acting. For further comments and an appropriate discussion of the formal content, see the second of Ref. [2].

$$\sqrt{\frac{\pi}{\alpha}}\sum_{r=0}^{\infty}\frac{1}{r!}\left(\frac{\hat{h}^2_{(\gamma,-\beta)}}{4\alpha}\right)^r = \\ = \sqrt{\frac{\pi}{\alpha}}\sum_{r=0}^{\infty}\frac{1}{r!}H_{2r}\left(\frac{\gamma}{2\sqrt{\alpha}},-\frac{\beta}{4\alpha}\right) = \sqrt{\frac{\pi}{\alpha+\beta}}e^{\frac{\gamma^2}{4(\alpha+\beta)}} \tag{7}$$

which is obtained after using the identity [2]

$$\sum_{r=0}^{\infty}\frac{t^r}{r!}H_{2r}(x,y) = \frac{1}{\sqrt{1-4yt}}e^{\frac{x^2 t}{1-4yt}} \tag{8}$$

The result in Eq. (7), evidently correct, yields some confidence on the reliability of the formalism, which is based on a rather wild use of the umbral formalism.

2 The Umbral Method and Generalized Hermite Polynomials

Even though not explicitly stated, the umbral operator defined in Eq. (5) satisfies the identity[2]

$$\hat{h}^m\hat{h}^r = \hat{h}^{m+r} \tag{9}$$

where the powers m and r are not necessarily real integers. It is also fairly natural to set

$$\partial_{\hat{h}}\hat{h}^r = r\,\hat{h}^{r-1} = rH_{r-1}(\gamma,-\beta) \tag{10}$$

Since the following recurrence holds

$$\partial_\gamma H_r(\gamma,-\beta) = rH_{r-1}(\gamma,-\beta) \tag{11}$$

the "derivative" operator can therefore be identified with

$$\partial_{\hat{h}} \to \partial_\gamma \tag{12}$$

Furthermore since

$$\hat{h}\hat{h}^r = \hat{h}^{r+1} = H_{r+1}(\gamma,-\beta) \tag{13}$$

and, on account of the recurrence,

$$H_{r+1}(\gamma,-\beta) = \gamma H_r(\gamma,-\beta) - 2\,\beta\,r\,H_{r-1}(\gamma,-\beta) \tag{14}$$

---

[2]The subscript $(\gamma,-\beta)$ has been omitted because the identity holds for $\hat{h}$ operators with the same basis, hereafter it will be included whenever necessary.

we can also conclude that $\hat{h}$ itself can be identified with the differential operator

$$\hat{h} = \gamma - 2\,\beta\,\partial_\gamma \tag{15}$$

It is also worth noting that

$$\begin{aligned} \partial_x^r e^{-\hat{h}x} &= (-1)^r \hat{h}^r e^{-\hat{h}x} = (-1)^r \sum_{n=0}^{\infty} \frac{(-x)^n}{n!} \hat{h}^{n+r} = \\ &= (-1)^r \sum_{n=0}^{\infty} \frac{(-x)^n}{n!} H_{n+r}(\gamma, -\beta) \end{aligned} \tag{16}$$

and, according to the identity

$$\sum_{n=0}^{\infty} \frac{t^n}{n!} H_{n+l}(x, y) = H_l(x + 2yt, y) e^{xt+yt^2} \tag{17}$$

we can establish the rule

$$\hat{h}^r e^{-\hat{h}x} = H_r(\gamma + 2\,\beta\,x, -\beta) e^{-(\gamma x + \beta x^2)} \tag{18}$$

We can now make a step further by defining the integral

$$\begin{aligned} I(\gamma, \beta) &= \int_{-\infty}^{\infty} e^{-\hat{h}x^2} dx \\ e^{-\hat{h}x^2} &= e^{-(\gamma x^2 + \beta x^4)} \end{aligned} \tag{19}$$

which, after applying the prescription that $\hat{h}$ can be treated as an ordinary algebraic quantity, writes

$$I(\gamma, \beta) = \sqrt{\pi} \hat{h}^{-\frac{1}{2}} \tag{20}$$

which makes sense only if we can provide a meaning for $\hat{h}^{-\frac{1}{2}}$, the most natural conclusion is that they can be understood as factional order Hermite, which for our purposes can be defined as it follows [4]

$$H_\nu(x, -y) = y^{\frac{\nu}{2}} \frac{e^{\frac{x^2}{4y}}}{\sqrt{\pi}} \int_0^{\infty} e^{-\frac{t^2}{4}} t^\nu \cos\left(\frac{x}{2\sqrt{y}} t - \frac{\pi}{2}\nu\right) dt \tag{21}$$

or as

$$H_\nu(x, -y) = \Gamma(\nu + 1) \sum_{r=0}^{\infty} \frac{x^{\nu-2r}(-y)^r}{\Gamma(\nu + 1 - 2r) r!}, \qquad x >> y \tag{22}$$

which has however a limited range of convergence.

The correctness of Eq. (20) can be readily proved by a numerical check, involving either the definitions (21) and (22).
We will comment, later in this paper, on the extension of the Hermite polynomials to non-integer index.

Let us now consider the following repeated derivatives

$$\begin{aligned} \partial_x^n e^{-\hat{h}x^2} &= (-1)^n H_n(2\,\hat{h}\,x, -\hat{h})e^{-\hat{h}x^2} = \\ &= (-1)^n n! \sum_{r=0}^{[\frac{n}{2}]} \frac{(-1)^r (2x)^{n-2r}}{(n-2r)!r!} \left(\hat{h}^{n-r} e^{-\hat{h}x^2}\right) \end{aligned} \tag{23}$$

Thus, getting, on account of Eq. (18),

$$\partial_x^n e^{-\hat{h}x^2} = (-1)^n n! \sum_{r=0}^{[\frac{n}{2}]} \frac{(-1)^r (2x)^{n-2r}}{(n-2r)!r!} H_{n-r}(\gamma + 2\,\beta x^2, -\beta) e^{-(\gamma x^2 + \beta x^4)} \tag{24}$$

in accordance with

$$\partial_x^n e^{-(\gamma x^2 + \beta x^4)} = H_n^{(4)}(-2\,\gamma x - 4\,\beta x^3, -\gamma - 6\,\beta x^2, -4\,\beta x, -\beta) e^{-(\gamma x^2 + \beta x^4)} \tag{25}$$

In Ref. [5] the following integral

$$\begin{aligned} J(a,b,c) &= \int_{-\infty}^{\infty} e^{-(ax^4+bx^2+cx)} dx, \\ Re(a) &> 0 \end{aligned} \tag{26}$$

has been considered, within the framework of problems regarding the non-perturbative treatment of the anharmonic oscillator. A possible perturbative treatment is that of setting

$$\begin{aligned} J(a,b,c) &= \sum_{n=0}^{\infty} \frac{(-1)^n}{n!} g_n(a) \left[H_n(c,-b) + H_n(-c,-b)\right], \\ g_n(a) &= \int_0^{\infty} x^n e^{-ax^4} dx = \frac{1}{4} a^{-\frac{n+1}{4}} \Gamma\left(\frac{n+1}{4}\right) \end{aligned} \tag{27}$$

which, as noted in [5], is an expansion with zero radius of convergence in spite of the fact that $J(a,b,c)$ is an entire function for any real or complex value of b, c.

The use of our point of view allows to write

$$J(a,b,c) = \int_{-\infty}^{\infty} e^{-\hat{h}_{(b,-a)}x^2 - cx} dx = \sqrt{\frac{\pi}{\hat{h}}} e^{\frac{c^2}{4\hat{h}}} = \sqrt{\pi} \sum_{s=0}^{\infty} \frac{1}{s!} \left(\frac{c}{2}\right)^{2s} \hat{h}^{-(s+\frac{1}{2})} \tag{28}$$

We have omitted the subscript $(b, -a)$ in the r.h.s. of Eq. (28) to avoid a cumbersome notation. The meaning of the operator $\hat{h}$ raised to a negative exponent is easily understood as

$$\hat{h}^{-\left(s+\frac{1}{2}\right)} = H_{-\left(s+\frac{1}{2}\right)}(b, -a) \tag{29}$$

where the negative index Hermite polynomials are expressed in terms of the parabolic cylinder functions D_n according to the identity [6]

$$H_{-n}(x, -y) = (2y)^{-\frac{n}{2}} e^{\frac{x^2}{8y}} D_{-n}\left(\frac{x}{\sqrt{2y}}\right) \tag{30}$$

The use of Eq. (30) in Eq. (28) finally yields the same series expansion obtained in Ref. [5]

$$J(a, b, c) = \sqrt{\pi} \sum_{s=0}^{\infty} \frac{1}{s!} \left(\frac{c}{2}\right)^{2s} (2a)^{-\frac{1}{2}\left(s+\frac{1}{2}\right)} e^{\frac{b^2}{8a}} D_{-\left(s+\frac{1}{2}\right)}\left(\frac{b}{\sqrt{2a}}\right) \tag{31}$$

which is convergent for any value of b, c and $a > 0$.

3 Final Comments

Regarding the use of non-integer Hermite polynomials it is evident that the definition adopted in Eq. (21) can be replaced by the use of the parabolic cylinder function, it is therefore worth noting that the use of the properties of the D functions allows the following alternative form for Eq. (20) (see Ref. [7])

$$\begin{aligned} I(\gamma, \beta) &= \sqrt{\pi}\, (2\,\beta)^{-\frac{1}{4}}\, e^{\frac{\gamma^2}{8\beta}} D_{-\frac{1}{2}}\left(\frac{\gamma}{\sqrt{2\beta}}\right) = \\ &= \sqrt{\frac{\gamma}{2\sqrt{2\beta}}}\, (2\,\beta)^{-\frac{1}{4}}\, e^{\frac{\gamma^2}{8\beta}} K_{\frac{1}{4}}\left(\frac{\gamma^2}{8\beta}\right); \\ D_{-\frac{1}{2}}(z) &= \sqrt{\frac{z}{2\pi}} K_{\frac{1}{4}}\left(\frac{1}{4} z^2\right) \end{aligned} \tag{32}$$

where $K_\nu(z)$ is a modified Bessel function of the second kind.

A further example of application of the method developed so far is provided by

$$\begin{aligned} \int_0^\infty e^{-(\beta x^{2n} + \gamma x^n)} dx &= \int_0^\infty e^{-\hat{h}_{(\gamma,-\beta)} x^n} dx = \frac{1}{n} \Gamma\left(\frac{1}{n}\right) \hat{h}_{(\gamma,-\beta)}^{-\frac{1}{n}} = \\ &= \frac{1}{n} \Gamma\left(\frac{1}{n}\right) (2\beta)^{-\frac{1}{2n}}\, e^{\frac{\gamma^2}{8\beta}} D_{-\frac{1}{n}}\left(\frac{\gamma}{\sqrt{2\beta}}\right) \end{aligned} \tag{33}$$

In a forthcoming more detailed note we will extend the method to cases involving higher order Hermite polynomials. Just to provide an idea of the extension of the technique, we note that the use of this family of polynomials allows to cast the integral in Eq. (28) in the form

$$J(a,b,c) = \int_{-\infty}^{\infty} e^{\,{}_4\hat{h}_{(-c,-a)}x - bx^2}\,dx = \sqrt{\frac{\pi}{b}}\,e^{\frac{\left({}_4\hat{h}_{(-c,-a)}\right)^2}{4b}} \tag{34}$$

where

$$\left({}_4\hat{h}_{(-c,-a)}\right)^n = H_n^{(4)}(-c,-a) = (-1)^n n! \sum_{r=0}^{\left[\frac{n}{4}\right]} \frac{c^{n-4r}(-a)^r}{(n-4r)!\,r!} \tag{35}$$

with $H_n^{(4)}(c,-a)$ being a fourth order Hermite Kampé de Fériét [8] polynomial.

The series expansion of the right hand side of Eq. (34) in terms of fourth order Hermite converges in a much more limited range than the series (31) and has been proposed to emphasize the possibilities of the method we have proposed so far.

According to our formalism the Pearcey integral, widely studied in optics, within the framework of diffraction problems [9], is easily reduced to a particular case of Eq. (26), namely

$$J(1,x,-iy) = \int_{-\infty}^{\infty} e^{-(t^4+xt^2)+iyt}\,dt = \sqrt{\frac{\pi}{\hat{h}_{(x,-1)}}}\,e^{-\frac{y^2}{4\hat{h}_{(x,-1)}}} \tag{36}$$

and can be expressed in terms of parabolic cylinder functions, as indicated before. It is perhaps worth stressing that, in the literature a converging series for the Pearcey integral is given in the form [10]

$$\begin{aligned} J(1,x,-iy) &= \int_{-\infty}^{\infty} e^{-t^4-\hat{h}_{(iy,-x)}t}\,dt = \\ &= \int_{0}^{\infty} e^{-t^4}\left(e^{\hat{h}_{(iy,-x)}t} + e^{\hat{h}_{(-iy,-x)}t}\right)dt = 2\sum_{n=0}^{\infty}(-1)^n g_{2n}(1)a_{2n}(x,y) \end{aligned} \tag{37}$$

with

$$\begin{aligned} a_0(x,y) &= 1, \\ a_1(x,y) &= y, \\ a_n(x,y) &= \frac{1}{n}\left(y\,a_{n-1}(x,y) + 2\,x\,a_{n-2}(x,y)\right) \end{aligned} \tag{38}$$

which is reconciled with our previous result, in terms of two variable Hermite polynomials, provided that one recognizes

$$J(1,x,-iy)=\sum_{n=0}^{\infty}\frac{(-1)^n}{n!}g_n(1)\left[H_n(-iy,-x)+H_n(iy,-x)\right] \tag{39}$$

In this letter we have provided some hint on the use of the Hermite calculus to study integral forms with specific application in different field of research. In a forthcoming more detailed note we will show how the method can be extended to a systematic investigation of the Voigt functions and to the relevant generalizations [11].

A further important application is the use of the method within the framework of evolutive PDE. To start, we consider the following straightforward example:

$$\begin{aligned}&\partial_t F(x,t)=\left(\alpha\partial_x+\beta\partial_x^2\right)F(x,t)\\&F(x,0)=f(x)\end{aligned} \tag{40}$$

whose (formal) solution is easily obtained as

$$F(x,t)=e^{(\alpha t)\partial_x+(\beta t)\partial_x^2}f(x) \tag{41}$$

The use of the formalism developed so far allows to write the rhs of Eq. (41) in the form

$$F(x,t)=e^{\hat{h}_{(\alpha t,\,\beta t)}\partial_x}f(x) \tag{42}$$

by the use of standard exponential rules we obtain

$$F(x,t)=f\left(x+\hat{h}_{(\alpha t,\,\beta t)}\right) \tag{43}$$

which is still a formal solution unless we provide a meaning for the rhs of Eq. (43). Let us therefore use the Fourier transform method to write

$$\begin{aligned}f\left(x+\hat{h}_{(\alpha t,\,\beta t)}\right)&=\frac{1}{\sqrt{2\pi}}\int_{-\infty}^{\infty}\tilde{f}(k)e^{ikx+\hat{h}_{(\alpha t,\,\beta t)}ik}dk=\\&=\frac{1}{\sqrt{2\pi}}\int_{-\infty}^{\infty}\tilde{f}(k)e^{ik(x+\alpha t)-k^2\beta t}dk\end{aligned} \tag{44}$$

which is a kind of Gabor transform [12]. It is evident that the same result can be obtained with ordinary means, we have used this example to prove the correctness and flexibility of the method we propose.

Let us now specialize the result to the case $f(x)=x^n$ and write

$$\begin{aligned}f\left(x+\hat{h}_{(\alpha t,\,\beta t)}\right)&=\left(x+\hat{h}_{(\alpha t,\,\beta t)}\right)^n=\sum_{s=0}^{n}\binom{n}{s}x^{n-s}\hat{h}^s_{(\alpha t,\,\beta t)}=\\&=H_n(x+\alpha t,\ \beta t)\end{aligned} \tag{45}$$

which is just the derivation from a different point of view of the following operational identity [13]

$$e^{\kappa\partial_x+\lambda\partial_x^2}x^n = H_n(x+\kappa,\lambda) \tag{46}$$

We want also to emphasize that the methods we have just outlined offer further possibility of speculation on the nature of the $\hat{h}$ operator itself, which can be further exploited to introduce nested Hermite polynomials. Its umbral nature and the associated flexibility allows indeed noticeable degree of freedom therefore the following identity derives naturally from the previous prescription

$$\begin{aligned} e^{\hat{h}_{(\alpha,\,\beta)}\partial_x^2}x^n &= H_n\left(x,\hat{h}_{(\alpha,\,\beta)}\right) = \\ &= n!\sum_{r=0}^{[\frac{n}{2}]}\frac{x^{n-2r}H_r(\alpha,\ \beta)}{(n-2r)!r!} \end{aligned} \tag{47}$$

This is what we define nested Hermite, for obvious reasons. The geometrical nature of the operator itself can be understood as discussed in [14] and will be further commented elsewhere.

Before closing the paper we want to underline that the possibilities for the applicability of the integration method discussed in this letter arise if, inside the integrand, an exponential generating function is recognized.

To clarify this point we note that the integral

$$\begin{aligned} &f(a,b,c) = \int_{-\infty}^{\infty} e^{-ax^2+\sqrt{x^2+bx+c}}dx, \\ &b^2-4c<0 \\ &a>1 \end{aligned} \tag{48}$$

can be written as

$$f(a,b,c) = \sqrt{\frac{\pi}{a}}e^{\frac{\hat{R}^2}{4a}} = \sqrt{\frac{\pi}{a}}\sum_{n=0}^{\infty}\frac{1}{n!}\left(\frac{1}{4a}\right)^n R_{2n}(b,c) \tag{49}$$

provided that

$$\begin{aligned} &e^{\sqrt{x^2+bx+c}} = e^{\hat{R}x} = \sum_{n=0}^{\infty}\frac{x^n}{n!}\hat{R}^n, \\ &\hat{R}^n = R_n(b,c) \end{aligned} \tag{50}$$

where $R_n(b,c)$ are polynomials of the parameter b, c.

Even though such a polynomials expansion can be obtained using different procedure, we have tested the validity of our ansatz using the following integral definition

$$R_m(b, c) = \frac{m!}{2\pi} \int_0^{2\pi} e^{-im\phi} e^{\sqrt{e^{2i\phi}+be^{i\phi}+c}} d\phi \tag{51}$$

which has been used to benchmark the identity (49), with the full numerical integration of (48). Further comments will be provided elsewhere.

References

1. E.W. Weisstein, *Hermite Number, from MathWorld–A Wolfram Web Resource*, http://mathworld.wolfram.com/HermiteNumber.html
2. G. Dattoli, B. Germano, M.R. Martinelli, P.E. Ricci, Lacunary generating functions of Hermite polynomials and symbolic methods. Ilirias J. Math. vol. 4(1), 16–23 (2015). ISSN: 2334-6574, http://www.ilirias.com; (G. Dattoli, E. Di Palma, E. Sabia, K. Gorska, A. Horzela, K. Penson *Operational Versus Umbral Methods and the Borel Transform* (2015). arXiv:1510.01204v1 [math.CA])
3. H.M. Srivastava, L. Manocha, A treatise On Generating Functions. Bull. Am. Math. Soc. (N.S.), vol. 19(1), 346–348 (1988)
4. R. Hermann, *Fractional Calculus: An Introduction for Physicists*, 2nd edn. (World Scientific, Singapore, 2014)
5. J. Bohacik, P. Augustin, P. Presnajder, Non-perturbative anharmonic correction to Mehler's presentation of the harmonic oscillator propagator. Ukr. J. Phys. **59**, 179 (2014)
6. M. Abramovitz, I.A. Stegun, (eds), *Parabolic Cylinder Function*, Chap. 19 in Handbook of Mathematical Functions with Formulas, Graphs and Mathematical Tables, 9th Printing (New York, Dover, 1972), pp. 685–700
7. E.W. Weisstein, *Parabolic Cylinder Function, from MathWorld–A Wolfram Web Resource*, http://mathworld.wolfram.com/ParabolicCylinderFunction.html
8. P. Appél, J. Kampé de Fériét, *Fonctions hypérgeométriques and Hypérspheriques; polinómes dHermite* (Gauthier-Villars, Paris, 1926)
9. J.L. López, P.J. Pagola, arXiv:1601.03615 [mat.NA]
10. M.V. Berry, C.J. Howls, Integrals with coalescing saddles, in *NIST Handbook of Mathematical Functions* (Cambridge University Press, Cambridge, 2010), pp. 775793. (Chapter 36)
11. M.A. Pathan, *Scientia*, Series A: Mathematical Sciences, vol. 12(9) (2006)
12. *Gabor Analysis and Algorithms Theory and Applications*, ed. By H.G. Feichtinger, T. Strohmer, Birkhuser (1998)
13. D. Babusci, G. Dattoli, M. Del Franco, *Lectures on Mathematical Methods for Physics*, RT/2010/58/ENEA (2010)
14. M. Artioli, G. Dattoli, *The Geometry of Hermite Polynomials*. Wolfram Demonstrations Project, Published: March 4 (2015), http://demonstrations.wolfram.com/TheGeometryOfHermitePolynomials/

Modal Epistemic Łukasiewicz Logic with Constant and Its Application in Immune System

Antonio Di Nola, Revaz Grigolia and Nunu Mitskevich

Abstract The theory of three-valued multi modal epistemic Łukasievicz logic with constant, which is an extension of the three-valued Łukasiewicz logic, the language of which is extended by nullary and unary connectives is developed in this paper. The unary connectives are interpreted as modal operators (knowledge operators). We propose to use such logic in studying immune system. A relational system is developed as a semantic of this logic. The relational systems represent the immune system which in its turn is a part of relational biology.

Keywords Agent · Many valued logic · Epistemic logic · Kripke model · Immune system

1 Introduction

We introduce a new logical system to study a biological system. The first attempt to study systems biology by means of logic (an axiomatic formal system) belongs to H. Woodger [27]. He proposed to discuss biology with precision of statements and reliability of reasoning. Here we also should to mention N. Rashevsky [20] and R. Rosen [21, 22]. N. Rashevsky founded relational biology that is the study of biology from the standpoint of definition of relations between the parts of a biological system. In this paper we analyze the responses from relational processes when they are represented by relational systems (Kripke frames) which are models of a such biological system as an immune system.

A. Di Nola (✉)
Department of Mathematics, University of Salerno, Salerno, Italy
e-mail: adinola@unisa.it

R. Grigolia
Department of Mathematics, Iv. Javakhisvili Tbilisi State University, Tbilisi, Georgia
e-mail: revaz.grigolia@tsu.ge; revaz.grigolia359@gmail.com

N. Mitskevich
Department of Biology, Iv. Javakhishvili Tbilisi State University, Tbilisi, Georgia
e-mail: nunu.mitskevichi@tsu.ge; nmitskevich@gmail.com

J. Gielis et al. (eds.), *Modeling in Mathematics*,
Atlantis Transactions in Geometry 2, DOI 10.2991/978-94-6239-261-8_5

In order to survive, all organisms must use energy sources present in the environment, and avoid dangers that could destroy them. To that end, they must acquire knowledge about the environment. All organisms acquire such knowledge, thanks to which they assume behaviors that, when successful, ensure their survival. Knowledge is a natural phenomenon that occurs in all organisms. Knowledge has a biological role, just like other capacities which ensure the survival of organisms.

Our investigation concerns to study mathematical objects - mathematical logical systems and their semantics - relational systems, with application for immune system. In turn the immune system consists of different kind of special cells - B and T lymphocytes. T cells and their subpopulations form some network like relational systems.

Our basic aim is to give to immunologists some useful tools for diagnosis about a state of immune system having some initial data. These data represent some properties of some parts of an immune system, that may be estimated, in particular some T cells being fundamental elements of the immune system.

Recent advances in Multiagent Systems and Epistemic Logic within Distributed Systems Theory, have used the Kripke model structure of models for the logic. In [19] examined one of the simpler versions of these models, interpreted systems, and the related Kripke semantics of the logic $S5_n$ (an epistemic logic with n agents).

This paper is a continuation of our previous work [7] where have been introduced multimodal epistemic three-valued Łukasiewicz logic $E\text{Ł}_3(n)$ which is used for estimation of the elements of relational system (Kripke frame) by the elements of adequate three element algebra corresponding to three-valued Łukasiewicz logic Ł$_3$. We introduce new logic - multimodal epistemic three-valued Łukasiewicz logic $E\text{Ł}_3^c(n)$ with constant. The logic $E\text{Ł}_3^c(n)$ is obtained from three-valued Łukasiewicz propositional logic Ł$_3$ by adding nullary connective c interpreted as "unknown" and n 'knowledge operators' ($n \geq 1$), with corresponding axioms. Notice that representing the connectives of this logic by functions (which are given in next section) on three-element set $\{0, \frac{1}{2}, 1\}$ we obtain complete system of functions, i.e. by means of the functions $x \oplus y = \min(1, x+y)$, $x \odot y = \max(0, x+y-1)$, $x^* = 1-x$ and constant $\frac{1}{2}$ we can express any function $f : \{0, \frac{1}{2}, 1\}^n \to \{0, \frac{1}{2}, 1\}$, in other words we represent Post functions. The knowledge operators model a community of ideal knowledge agents who have the properties of veridical knowledge (everything they know is true), fuzzy knowledge (everything they know is quasi true, positive introspection (they know what they know) and negative introspection (they know what they do not know) and so on. The knowledge operators permit the following interpretation:

$\Box_i \alpha$ - "the agent i knows proposition α";

$\Diamond_i \alpha$ - "the agent i does not know that proposition α is false".

For detail information on classical and non-classical modal logic we refer to [1–4, 8, 9, 13, 14, 17–19, 25].

We will show that 3-valued multimodal Łukasiewicz logic $E\text{Ł}_3^c(n)$ is complete with respect to descriptive Kripke frames and use the ones for representation of immune system. The set of T-cells we can understand as the set of agents. In many ways the immune system is a black box; although many of its inputs and outputs are known, exactly how the system achieves its function is the subject of many investigations. Laboratory experiments provide large quantities of data, allowing components (agents (T-cells)) within the black box to be identified, but there remain many details of how the components, (state of the agents (T-cells)) of the system carry out their functions, or on the nature of interaction between components. There are many variables in such systems that exhaustive testing to establish these details is not feasible. Multimodal logic and corresponding to it Kripke model are ideally suited to describing immune system at this level: they may be represented as a relational system of interacting elements (components), where the components themselves may have complex, non-deterministic, individual behaviour. Moreover, the usage of multimodal logic and Kripke model gives access to a range of investigatory techniques, including simulation, verification via logical properties.

Let us remark that if $n = 1$, then $EŁ_3(1)$ coincides with the monadic propositional logic [6], i.e. the modal operator behaves as a quantifier. In other words a monadic propositional logic is connected with the first-order logic with fixed one individual variable x. More precisely, let L denote a first-order language based on $\cdot, +, \rightarrow, \neg, \exists$ and let L_m denote a propositional language based on $\cdot, +, \rightarrow, \neg, \exists$. Let $Form(L)$ and $Form(L_m)$ be the set of all formulas of L and L_m, respectively. We fix a variable x in L, associate with each propositional letter p in L_m a unique monadic predicate $p^*(x)$ in L and define by induction a translation $\Psi : Form(L_m) \rightarrow Form(L)$ by putting:

- $\Psi(p) = p^*(x)$ if p is propositional variable,
- $\Psi(\alpha \circ \beta) = \Psi(\alpha) \circ \Psi(\beta)$, where $\circ = \cdot, +, \rightarrow$,
- $\Psi(\exists\alpha) = \exists x \Psi(\alpha)$.

Through this translation Ψ, we can identify the formulas of L_m with monadic formulas of L containing the variable x.

Having as a domain of interpretation some set of T-cells, a monadic predicate is interpreted as some property of a T-cell, say, for example, to possess some geometrical shape. This reasoning can be generalized on $EŁ_3^c(n)$.

2 Multimodal Epistemic 3-Valued Łukasiewicz Logics $EŁ_3^c(n)$

The unit interval of real numbers [0, 1] endowed with the following operations: $x \oplus y = \min(1, x + y)$, $x \odot y = \max(0, x + y - 1)$, $x^* = 1 - x$, becomes an MV-algebra [5]. It is well known that the MV-algebra $S = ([0, 1], \oplus, \odot, ^*, 0, 1)$ generate the variety **MV** of all MV-algebras, i.e. $\mathcal{V}(S) = \mathbf{MV}$.

Let Q denote the set of rational numbers, for $(0 \neq) n \in \omega$ we set

$$S_n = (S_n; \oplus, \odot, ^*, 0, 1),$$

where

$$S_n = \left\{0, \frac{1}{n}, \ldots, \frac{n-1}{n}, 1\right\}.$$

For any positive integer n, S_n is a subalgebra of S. We are interested by $S_2 = (S_2; \oplus, \odot, ^*, 0, 1)$.

The formulas of Łukasiewicz logic Ł are built from a countable set of propositional variables $Var = \{p, q, ...\}$ using the connectives & (strong conjunction), $\to$ (implication) and $\bot$ (falsity truth constant).

In order to introduce the infinite-valued Łukasiewicz logic we start by considering the standard MV-algebra $S = ([0, 1], \odot, \Rightarrow, 0)$ (which is functionally equivalent to the MV-algebra defined above), where a binary operation $\odot$ called Łukasiewicz t-norm and defined as $a \odot b = max\{0, a + b - 1\}$, for all $a, b \in [0, 1]$ [12]; a binary operation $\Rightarrow$ called the residuum (of the t-norm $\odot$) and defined as $a \Rightarrow b = min\{1, 1 - a + b\}$, and $\neg a = a \Rightarrow 0 = 1 - a$, $a \oplus b = \neg(\neg a \odot \neg b) = min(1, a + b)$, for all $a, b \in [0, 1]$.

We extend 3-valued Łukasiewicz logic $Ł_3$ to the 3-valued multimodal Łukasiewicz logic $EŁ_3^c(n)$ by adding nullary connective c and n unary modal operators $\Box_i$ and $\Diamond_i$ $(i = 1, \ldots, n)$ to the language of $Ł_3$. The formulas of 3-valued multimodal Łukasiewicz logic $EŁ_3^c(n)$ are built from a countable set of propositional variables $Var = \{p, q, \ldots\}$ using the connectives & (strong conjunction), $\to$ (implication) and $\bot$ (falsity truth constant), c (unknown truth constant), and $2n$ unary modal operators $\Box_i$ and $\Diamond_i$ $(i = 1, \ldots, n)$ in usual way. Denote by $Ł_3^c$ the three-valued propositional Łukasiewicz logic with constant in the language $Var = \{p, q, \ldots\}$ using the connectives & (strong conjunction), $\to$ (implication) and $\bot$ (falsity truth constant), c (unknown truth constant).

A *propositional evaluation* is a homomorphism e from the algebra of formulas into the algebra S_2, i.e., a mapping e from the set of formulas into $\{0, \frac{1}{2}, 1\}$ such that

- $e(\varphi\&\psi) = e(\varphi) \odot e(\psi)$,
- $e(\varphi \to \psi) = e(\varphi) \Rightarrow e(\psi)$,
- $e(\bot) = 0$,
- $e(c) = \frac{1}{2}$.

A formula φ is said to be *valid* when it is evaluated to 1 in all propositional evaluations. Then, three-valued Łukasiewicz logic $Ł_3^c$ is defined as the set of valid formulas.

We introduce the connectives $\wedge$, $\vee$, $\leftrightarrow$, $\neg$, $\underline{\vee}$ and $\top$ (the semantics counterpart will be denoted, respectively, by $\wedge$, $\vee$, $\Leftrightarrow$, $\neg$, $\oplus$ and 1, and $\odot$ for &) as the following abbreviations: $\varphi \wedge \psi = \varphi\&(\varphi \to \psi)$, $\varphi \vee \psi = (\varphi \to \psi) \to \psi$, $\varphi \leftrightarrow \psi = (\varphi \to \psi)\&(\psi \to \varphi)$, $\neg\varphi = \varphi \to \bot$, $\varphi\underline{\vee}\psi = \neg(\neg\varphi\&\neg\psi)$, $\top = \neg\bot$.

The logic Ł is axiomatized by the following axiom schemata:

Ł1. $\varphi \to (\psi \to \varphi)$,
Ł2. $(\varphi \to \psi) \to ((\psi \to \chi) \to (\varphi \to \chi))$,

Ł3. $((\varphi \to \psi) \to \psi) \to ((\psi \to \varphi) \to \varphi)$,
Ł4. $(\neg\varphi \to \neg\psi) \to (\psi \to \varphi)$.

The inference rule is Modus Ponnens: $\varphi, \varphi \to \psi/\psi$.

3-valued Łukasiewicz logic $Ł_3$ is axiomatized by the axioms of Ł plus the schema: Ł5. $(\varphi\&\varphi) \leftrightarrow (\varphi\&\varphi\&\varphi)$. 3-valued Łukasiewicz logic $Ł_3^c$ is axiomatized by the axioms of $Ł_3$ plus the schemata: C1. $(c\&c) \leftrightarrow \bot$, C2. $(c\underline{\vee}c) \leftrightarrow \top$, C3. $(c \wedge \top) \leftrightarrow c$, C4. $(c \wedge \bot) \leftrightarrow \bot$, C5. $\neg c \leftrightarrow c$.

2.1 3-Valued Descriptive Kripke Models

A *3-valued Kripke frame for agent i* is a pair $\mathfrak{J}_i = (W_i, R_i)$, $i = 1, \ldots, n$, consisting of a non-empty set W_i of elements called *the states of the agent i* (or *possible worlds of the agent i*); $R_i \subset W_i \times W_i$ is a binary reflexive and transitive relation on W_i (called *the accessibility relation for agent i*).

A 3-*valued Kripke model* for agent i (or simply, Kripke model for agent i, when there is no ambiguity) is a pair $\mathfrak{M}_i = (\mathfrak{J}_i, e_i)$, $i = 1, \ldots, n$, where $\mathfrak{J}_i = (W_i, R_i)$ is Kripke frame for agent i and $e_i : Var \times W_i \to S_2$ is a function, called *evaluation for agent i*, which maps every propositional variable $p \in Var$ and possible world $w \in W_i$ to the set of truth values S_2, $i = 1, \ldots, n$, such that if $e_i(p, w) = 1$ and $(w, w') \in R_i$ then $e_i(p, w') = 1$. If φ is a propositional formula of $Ł_3$, then $e_i(\varphi, w) \in S_2$ is a propositional evaluation for agent i; if φ is a modal formula, then $e_i(\Diamond_i\varphi, w) = \bigvee\{e_i(\varphi, w\prime) : (w, w\prime) \in R_i\}$; $e_i(\Box_i\varphi, w) = \bigwedge\{e_i(\varphi, w\prime) : (w, w\prime) \in R_i\}$ for every $w \in W_i$, $i = 1, \ldots, n$.

A modal formula φ is said to be *modally valid for agent i* when it is evaluated to 1 in all Kripke models for agent i; it is said to be *modally* 1-*satisfiable for agent i* when there is some Kripke model for agent i and some world w such that $e_i(\varphi, w) = 1$; and it is said to be *modally satisfiable for agent i* when it is 1-satisfiable for agent i. A modal formula φ is said to be *modally valid* when it is evaluated to 1 in all Kripke models for every agent i.

A 3-*valued descriptive Kripke frame* is a pair $\mathfrak{J} = (W, R)$, $W = \{W_1, \ldots, W_n\}$ is the set of n agents (or *possible worlds*); $R \subset W \times W$ is a binary reflexive and transitive relation on W (called *the accessibility relation between agents* $i(= W_i)$).

A 3-*valued descriptive Kripke global model* (or *descriptive Kripke global model*) is a triple $\mathfrak{M} = (W, R, V)$ where $W = \{W_1, \ldots, W_n\}$ is the set of n agents (or *possible worlds*); $R \subset W \times W$ is a binary relation on W (called *the accessibility relation between agents* $i(= W_i)$); $V(\varphi, W_i) = \bigwedge\{e_i(\varphi, w) : w \in W_i, e_i : Var \times W_i \to S_2\}$, $V(\Box\varphi, W_i) = \bigwedge\{V(\varphi, W_j) : (W_i, W_j) \in R\}$, $V(\Diamond\varphi, W_i) = \bigvee\{V(\varphi, W_j) : (W_i, W_j) \in R\}$, where $\Box\varphi$ and $\Diamond\varphi$ are a abbreviations of $\Box_1\varphi \wedge \ldots \wedge \Box_n\varphi$ and $\Diamond_1\varphi \vee \ldots \vee \Diamond_n\varphi$ respectively.

A modal formula φ is said to be *globally modally valid* when it is evaluated to 1 in all Kripke models for every agent $i \in \{1, \ldots, n\}$; it is said to be *modally satisfiable* when it is 1-satisfiable for some agent $i \in \{1, \ldots, n\}$.

The logic $E\text{Ł}_3^c(n)$ is defined as the set of its modal formulas that are globally modally valid. It is worth pointing out that for this modal logic the modal operators are interdefinable by means of the modally valid formulas $\Diamond_i\varphi \leftrightarrow \neg\Box_i\neg\varphi$ and $\Box_i\varphi \leftrightarrow \neg\Diamond_i\neg\varphi$.

2.2 Axiomatization

We suggest the following schemata of axioms for $E\text{Ł}_3^c(n)$: to the schemata of axioms of Ł_3^c we add

(1) $\Box_i\varphi \to \varphi, \quad i = 1, \ldots, n,$
(2) $\Box_i\varphi \to \Box_i\Box_i\varphi, \quad i = 1, \ldots, n,$
(3) $\Box_i(\varphi \wedge \psi) \leftrightarrow (\Box_i\varphi \wedge \Box_i\psi), \quad i = 1, \ldots, n,$
(4) $\Box_i(\varphi \& \varphi) \leftrightarrow (\Box_i\varphi \& \Box_i\varphi), \quad i = 1, \ldots, n,$
(5) $\Box_i(\varphi \underline{\vee} \varphi) \leftrightarrow (\Box_i\varphi \underline{\vee} \Box_i\varphi), \quad i = 1, \ldots, n,$
(6) $\Diamond_i\varphi \to \Box_i\Diamond_i\varphi, \; i = 1, \ldots, n,$

inference rules: $\varphi, \varphi \to \psi/\psi, \; \varphi/\Box_i\varphi, \; i = 1, \ldots, n.$

Notice that mono-modal fragment of the logic $E\text{Ł}_3(1)$ coincides with monadic three-valued Łukasiewicz logic [6, 23]. Extending this result to the $E\text{Ł}_3^c(1)$ we easily obtain three-valued monadic Post logic.

Remark. Notice that the algebra $S_2^c = (\{0, \frac{1}{2}, 1\}, \odot, \Rightarrow, 0, \frac{1}{2}, 1)$ is functionally complete, i.e. any function $f : \{0, \frac{1}{2}, 1\}^n \to \{0, \frac{1}{2}, 1\}$ is expressed by some term on the language $\odot, \Rightarrow, 0, \frac{1}{2}, 1$. In other words it is a Post algebra of order 3. Indeed, it is enough to express the operation $\sim\frac{x}{2} = \frac{x-1(mod\ 3)}{2}$ in S_2^c, where $x = 0, 1, 2$: $\sim\frac{x}{2} = \frac{1}{2} \odot \frac{x}{2} \vee \neg(\frac{x}{2} \oplus \frac{x}{2})$. It is known that any finite Post algebra is isomorphic to a finite product of algebras S_2^c. Like in monadic Boolean case a finite subdirectly irreducible monadic Post algebra is an algebra $(A, \Box)$, where A is a finite Post algebra and monadic operator $\Box$ is defined in the following way: $\Box(x_1, \ldots, x_n) = x_m$ where $(x_1, \ldots, x_n) \in A$ and $x_m = min\{x_1, \ldots, x_n\}$. The Post monadic logic is defined as all the formulas of $E\text{Ł}_3^c(1)$ that are valid in all finite monadic Post algebras $(A, \Box)$.

It is well known that the category of Boolean algebras and Boolean homomorphism is equivalent to the category of Post algebras and Post homomorphisms. It is well known that the category of monadic Boolean algebras and monadic Boolean homomorphisms is dually equivalent to the category of $S5$-frames and p-morphisms [4], where $S5$-frame is a Kripke frame (W, R) with equivalence relation R on W. So, we can conclude that the category of monadic Post algebras and monadic Post homomorphisms is dually equivalent to the category of $S5$-frames and p-morphisms. Hence, we have another semantical approach for $E\text{Ł}_3^c(1)$ taking $S5$-frames as 3-valued descriptive Kripke frame. And, following to [19], we can transfer this technic to the multi-modal logic $E\text{Ł}_3^c(n)$.

According to G. Hansoul and B. Teheux's results [13] about completeness theorem of mono-modal Łukasiewicz logic with respect to $n + 1$-Kripke completeness and adopting the ones to the multi-modal logic $EŁ_3^c(n)$ we arrive to the following completeness theorem

Theorem 1 *A formula of* $EŁ_3^c(n)$ φ *is a theorem of* $EŁ_3^c(n)$ *iff* φ *is globally valid.*

3 On the Cells of Immune System

The immune system represents a multi-component system which includes a range of cell types with different roles in defending the body against infection agents, damaged tissues and in preventing the uncontrolled growth of rogue cells or cancerous cells. Many of these cells arise in the bone marrow, circulate in the blood and can migrate into solid tissues. Immune responses involve interactions between some of these cells and/or their secreted products. B and T lymphocytes specifically react to microbial antigens: activated B lymphocytes secrete antigenbinding antibodies, and subpopulations of T lymphocytes possess helper, regulatory or cytotoxic functions. Natural killer cells are also cytotoxic cells of the lymphoid lineage, but they do not possess properties of antigen recognition. Different types of cells are antigen presenting cells (APC) present antigens to T lymphocytes. Blood monocytes give rise to tissue macrophages that are phagocytes, as are circulating neutrophils, which are the most plentiful type of granulocytes. Other circulating granulocytes are eosinophils that secrete toxic mediators, and basophils that, in common with tissue mast cells, are important sources of inflammatory mediators. Other cells contribute to immune and inflammatory responses, including endothelial cells, erythrocytes and platelets.

A variety of cell types are important components of the immune system. The main responsible cells are B and T lymphocytes with high specificity. They specifically recognise antigens and are responsible for adaptive, acquired immunity. B lymphocytes recognise native, unprocessed antigens via surface immunoglobulins and produce secreted immunoglobulins called antibodies. T lymphocytes recognise processed antigens, usually peptides associated with MHC proteins expressed on the surface of antigenpresenting cells. Different T lymphocyte subpopulations make a multi-functional network which will be described in this section more detail. T cells can be categorized based upon cell surface expression of either cluster of differentiation T4 (CD4) or T8 (CD8). CD4+ T cells recognize antigen presented in the context of class II major histocompatibility complex (MHC), while CD8+ cells recognize antigen presented in the context of class I MHC. CD4+ T helper subsets include T helper type 1 (Th1), T helper type 2 (Th2), and T helper type 17 (Th17) cells and some other new described types of Th cells [15, 16].

There is evidence that each of these subsets is involved in the defense against a certain subset of microorganisms. Th1 are pivotal in defense against intracellular microorganisms in general and mycobacteria in particular. Patients with mutations in the interferon-γ (IFN-γ) receptor or interleukin-12 (IL-12) receptor present with recurrent infections with mycobacteria and Salmonella. Th2 cells are integral in expelling parasitic infestations. Th17 seem to play a significant role in defense against extracellular bacteria and some fungi. Th1 and Th17 cells play major roles in autoimmunity, whereas Th2 cells are the hallmark of atopic disease. T regulatory (Treg) cells represent a major subset of CD4+ T cells that may be involved in regulating and attenuating the activity of the three T helper subsets.

T lymphocytes are a major source of cytokines. Cytokines are the messengers responsible for most of the biological effects in the immune system, such as cell-mediated immunity and antibody depended type responses.

Th1 cells are characterized by the production of pro-inflammatory cytokines. Th1 cells are involved in cell-mediated immunity. The cytokines produced by Th1 cells stimulate the phagocytosis and destruction of microbial pathogens. Several chronic inflammatory diseases have been described as Th1 dominant diseases. Th2 cells are thought to play a role in allergy responses. Atrophy and allergy are thought to be Th2 dominant conditions [24]. Improved understanding of Th1 and Th2 differentiation will improve our overall understanding of the immune system, its response to infection and aberrant responses that lead to disease.

The immune balance controlled by T helper 1 (Th1) and T helper 2 (Th2) is crucial for immunoregulation and its imbalance causes various immune diseases. Therefore, diagnosis of Th1/Th2 balance in autoimmune and atopic diseases including asthma is essential for the application of immune balance regulating drugs. Th1/Th2 balance is not only controlled by Th1 cells and Th2 cells, but also by various regulatory factors including regulatory T cells, sexual factor, chemokines, transcription factors, signal transduction pathway etc. Current research strategies seek to describe these multi-factorial system responsible to keep the Th1/Th2 balance in the body and predict some logical novel targets for regulating this balance which is based on another subpopulations of T cells - T regulatory (T reg) cells and their cytokine profiles. Naturally occurring CD4+ T regulatory cells (nTreg) are derived centrally in the thymus and constitutively express CD25 and other suppressive molecules including CTLA-4. These cells generally appear to exert suppressive effects by direct cell contact rather than cytokine production. The Foxp3 gene appears to be a critical regulator of the development of this subgroup of CD4+CD25+ nTreg cells. At a population level, there has been a parallel rise in both Th1-mediated autoimmune diseases and Th2-mediated allergic diseases. At the individual level, there is accumulating evidence that atopy is associated with an increase in both Th1 and Th2 responses. Furthermore, Th1 cells also appear to play a role in allergic inflammation in local tissues, failing to counter balance Th2 responses in airways inflammation. These observations lead to the opinion that the autoimmune diseases may develop as a result of a more fundamental failure of underlying immune regulation, rather than a simple skewing of immune response along a Th1/Th2 homeostasis as previously thought [10, 11].

The nonspecific part of the immune system and different cells involved in the first line defense are also very important components of this system: Natural killer (NK) cells are lymphoid cells that lack antigenspecific receptors, but mediate cytotoxic activity against infected or malignant cells. Dendritic cells (DC) are potent antigenpresenting cells involved in the activation of T lymphocytes, whereas follicular DC present antibodyassociated antigens to B lymphocytes. Monocytes are circulating blood cells that give rise to tissue macrophages with phagocytic and antigen presentation functions. Granulocytes circulate in the blood, migrate into tissues, and include phagocytic neutrophils and eosinophils that secrete toxic mediators, and basophils that release inflammatory mediators; mast cells are tissue cells with similar properties to basophils. A variety of other cell types contribute to the generation and regulation of immune and inflammatory responses, including endothelial cells, erythrocytes and platelets. This multifactorial and multifunctional immune system represents the relational biological system which will be described in this paper by using Modal Lukasiewicz Epistemic Logic.

4 Description of Immune System by Descriptive Kripke Frames

In this section we try to represent some simple fragments of an immune system by 3-valued Kripke frame with the following interpretation in immune models. We will consider two cases.

I. Let $\mathfrak{J}_1=(W_1, R_1)$ be 3-valued Kripke frame for agent 1, where $W_1=\{MPh, Ag_1, T_c, Th_0, Th_1\}$ and R_1 is the transitive closure of the binary relation $\{(Th_1, MPh), (MPh, Ag_1), (MPh, MPh), (T_c, Ag_1), (Ag_1, T_c), (Th_1, Tc), (Th_0, Th_1), (Ag_1, Th_0), (Ag_1, Ag_1), (Tc, Tc), (Th_1, Th_1), (Th_0, Th_0)\}$; $\mathfrak{J}_2 = (W_2, R_2)$ be 3-valued Kripke frame for agent 2, where $W_2 = \{Ag_2, Th_0, Th_2, B, Y_{Ab}\}$ and R_2 is the transitive closure of the binary relation $\{(Ag_2, Th_0), (Th_0, T2), (Th_2, B), (B, Y_{Ab}), (Ag_2, Ag_2), (B, B), (Th_2, Th_2), (Th_0, Th_0), (Y_{Ab}, Y_{Ab})\}$.

Let $W' = \{W_1, W_2\}$ be system of agents, which is the set of immune system. $\mathfrak{J}' = (W', R')$ is a global immune system represented as a global 3-valued descriptive Kripke frame, where $R' = \{(W_1, W_1), (W_2, W_2)\}$, see Fig. 1. Notice, that $\mathfrak{J}' = (W', R')$ represents a Kripke frame for the classical epistemic multimodal system $S5_2$.

II. Let $W'' = \{W_1, W_2\}$ be another system of agents and $\mathfrak{J}'' = (W'', R'')$ be a global immune system represented as a global 3-valued descriptive Kripke frame, where $R'' = \{(W_1, W_1), (W_2, W_2), (W_2, W_1)\}$, see Fig. 2.

Now we will give some interpretation in models of immune system. Let $\mathfrak{J} = (W, R)$, $W = \{W_1, \dots, W_n\}$ is the set of n agents (or *possible worlds*); $R \subset W \times W$ is a binary reflexive and transitive relation on W (called *the accessibility relation between agents* $i(= W_i)$).

A 3-*valued descriptive Kripke global model* (or *descriptive Kripke global model*) is a pair $\mathfrak{M} = (J, V)$ where $V(\varphi, W_i) = \bigwedge\{e_i(\varphi, w) : w \in W_i, e_i : Var \times W_i \rightarrow S_2\}$, $V(\Box\varphi, W_i) = \bigwedge\{V(\varphi, W_j) : (W_i, W_j) \in R\}$, $V(\Diamond\varphi, W_i) = \bigvee\{V(\varphi, W_j) : (W_i, W_j) \in R\}$, where $\Box\varphi$ and $\Diamond\varphi$ are a abbreviations of $\Box_1\varphi \wedge ... \wedge \Box_n\varphi$ and $\Diamond_1\varphi \vee ... \vee \Diamond_n\varphi$ respectively.

Fig. 1 Kripke frame $\Im' = (W', R')$

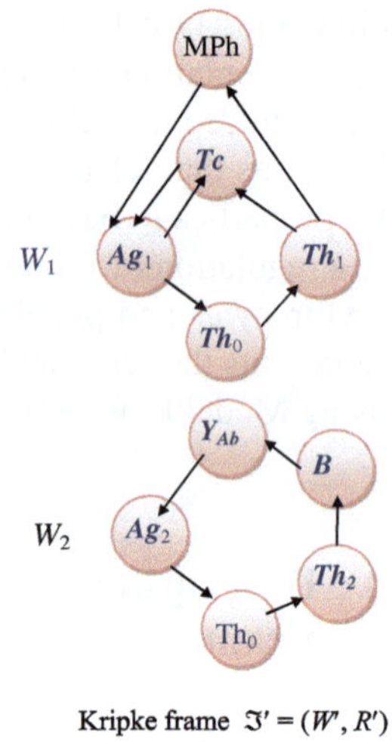

Fig. 2 Kripke frame $\Im'' = (W'', R'')$

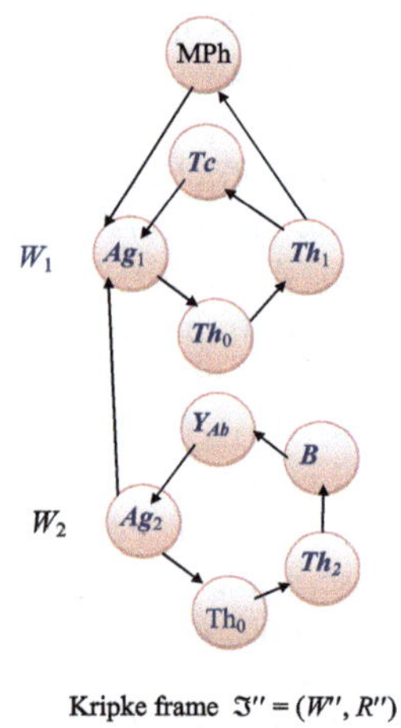

Recall that a modal formula φ is said to be modally valid for agent i when it is evaluated to 1 in all Kripke models for agent i; a modal formula φ is said to be globally modally valid when it is evaluated to 1 in all Kripke models for every agent $i \in \{1, \ldots, n\}$.

Now we give a naive definition of (global) immune system *ImS*. A (global) immune system *ImS* is a set of *Tcells* with some actions between them. Identifying *Tcell* with agent (or possible word) and an action between *Tcells* with the relation between agents we can represent a (global) immune system *ImS* as a (global) 3-valued descriptive Kripke frame.

4.1 The Epistemic Aspects of Immune System

We say that $w \in W_i$ is activated if $e_i(p, w) = 1$, it is not activated if $e_i(p, w) = 0$, it is not known that w is activated if $e_i(p, w) = 1/2$. So, for evaluation V we have the set of points of $\bigcup_{i=1}^n W_i$ such that part of them is activated, part of them is not activated and part of them is unknown they are activated or not activated.

A function $Es : \bigcup_{i=1}^n W_i \to S_2^c$ is named *an epistemic state* if for every $w, w' \in W_i$ it is hold

$$(w, w') \in R_i \Rightarrow (Es(w) = 1 \Rightarrow Es(w') = 1).$$

Let $e_i : Var \times W_i \to S_2^c$ be an evaluation for agent i, $i = 1, \ldots, n$. Let $e : Var \times \bigcup_{i=1}^n W_i \to S_2^c$ be an evaluation of $\bigcup_{i=1}^n W_i$ $(= W)$ such that $e(p, w) = e_i(p, w)$ if $w \in W_i$. A formula φ defines a function $S_\varphi^e(w) : \bigcup_{i=1}^n W_i \to S_2^c$ such that $S_\varphi^e(w) = e_i(\varphi, w)$ for $w \in W_i$. We say that a formula φ is *labelled by the evaluation e* if $S_\varphi^e(w)$ is an epistemic function and denote such kind of function by ES_φ^e. The process of transformation *Act* of one epistemic function Es_1 to an another epistemic function Es_2 we name "$\varphi - activation''$. So, for a formula φ a transferring of the epistemic state function ES_φ^e to the epistemic state function $ES_\varphi^{e'}$ is a φ-activation of points of $\bigcup_{i=1}^n W_i$ $(= W)$.

We described an immune system as a Kripke Frame. It means that by Kripke frame we capture just the relational structure of an immune system.

This representation of immune system neglects the epistemic information about the immune system, that is some knowledge on the points w are not represented. So to recover such an information we give the notion of *Epistemic State function* of an immune system. This is done by a function *Es* defined on all possible worlds to S_2^c. Of course *Es* satisfies some suitable conditions, which are essentially compatibility conditions with respect the relational structure of the immune system. In this way we have a more faithful representation of the knowledge about the given immune system. It is reasonable to think that to get the value $Es(w)$ it is needed some laboratory job. We plan mathematically to study the set of all epistemic states. Our aim is to help the immunologist to have a formal and canonical way to explore the possible Epistemic State (function) of an immune system. Since to an immune system, as defined in the paper, can be associated a logic which is complete with respect to certain Kripke frames, and since immune system representation gives us as Kripke Frame, we use formulas of the logic of our Kripke Frame immune system, to define some Epistemic States of the immune system. Actually we use a formula φ and an evaluation e of φ, in the following way: $ES_\varphi^e(w) = e(\varphi, w)$.

It is worth to note that a single formula φ essentially represents a set of Epistemic States, actually all such states defined by ES_φ^e when e varies in the set of all evaluations. In this way a given formula represents a collection of Epistemic States of the immune system. It could be of interest to explore the possibility to check whether given a collection of Epistemic States we can found a formula representing such a collection.

We defined the Activation function *Act* as a functions defined on the set of all Epistemic States with value in the same set. This is a way to represent how changes the epistemic information after, say an experiment, that produces new information about the epistemic values of all points w. To know facts about the function *Act* means to know facts about possible variations of the epistemic state of the system, and to check whether these variations can be described by formulas.

5 Conclusion and Perspectives

We have introduced the new logic - three valued multi-modal epistemic Łukasiewicz logic $EŁ_3^c(n)$ adequate semantic of which are special relational systems named 3-valued descriptive Kripke frames. In other words we have proven completeness theorem, i.e. a formula α of $EŁ_3^c(n)$ which is a theorem of $EŁ_3^c(n)$ iff it is (globally) modally valid. So we have shown that some immunological systems, described as relational systems, are Kripke Models of the logic we have presented. This means that theorems in our logic describe true properties in the model of immune system - the network of T lymphocytes' subpopulations and that we can set some conjecture on the model, which at the moment is not clear to medical and life sciences, and try to prove it as a theorem, or disprove it.

The principal role of the immune system is thought to be host defense against invasion by pathogenic agents. For this reason, the study of the immunology of infection offers important insight concerning effector functions and regulatory interactions fundamental to the immune response. In reacting to infectious agents, the immune system can generate to varying degrees unwanted immunopathologic side effects in the form of fever, tissue damage and immune complex lesions. The balance between resistance and pathology is delicate and determined both by the character of the agent and the immunoregulatory state of the host. It has become increasingly clear that cytokines, and in particular those associated with the Th1/Th2 CD4+ T cell subsets, are key determinants of the beneficial vs disease consequence of the host response [26].

So, if we have a theorem φ (which is a proposition) of multimodal epistemic 3-valued Łukasiewicz logic with constant $EŁ_3^c(n)$, then any evaluation in any immune system, represented as a 3-valued Kripke model, is equal to 1.

We have represented an immune system *ImS* as a relational system with epistemic information. This epistemic information is expressed by epistemic state function *Es*. Moreover, having some formula φ we define a collection of epistemic states which determined by the epistemic function ES_{φ}^{e}. In the perspective it will be interesting to find a corresponding formula for some epistemic state which is very important for immune system.

References

1. P. Blackburn, M. de Rijke, Y. Venema, *Modal Logic*. Cambridge Tracts in Theoretical Computer Science, vol. 53 (Cambridge University Press, Cambridge, 2001)
2. F. Bou, F. Esteva, L. Godo, R. Rodríguez, *n*-Łukasiewicz modal logic. Manuscript (2007)
3. X. Caicedo, R. Rodríguez, A Gödel similaritybased modal logic. Manuscript. A shortened version was published as A Gödel modal logic, in *Proceedings of Logic, Computability and Randomness 2004. Cordoba, Argentina* (2007)
4. A. Chagrov, M. Zakharyaschev, *Modal Logic*. Oxford Logic Guides, vol. 35. (Oxford University Press, Oxford, 1997)
5. C.C. Chang, Algebraic analysis of many-valued logics. Trans. Am. Math. Soc. **88**, 467–490 (1958)
6. A. Di Nola, R. Grigolia, On monadic MV-algebras. APAL **128**(1–3), 125–139 (2004)
7. A. Di Nola, R. Grigolia, N. Mitskevich, Multimodal epistemic lukasiewicz logics with application in immune system. Soft Comput. **19**(11), 3341–3351 (2015). doi:10.1007/s00500-015-1804-4
8. M. Fitting, Many-valued modal logics. Fundam. Inform. **15**, 235254 (1992)
9. M. Fitting, Many-valued modal logics, II. Fundam. Inform. **17**, 5573 (1992)
10. S. Fu, N. Zhang, A.C. Yopp, D. Chen, et al., TGF-β induces Foxp3 + T-regulatory cells from CD4 + CD25- precursors. Am. J. Transpl. **4**, 1614–1627 (2004)
11. S. Hori, S. Sakaguchi, Foxp3: a critical regulator of the development and function of regulatory T cells. Microbes Infect. **6**, 745–751 (2004)
12. P. Hajek, *Metamathematics of Fuzzy Logic* (Kluwer Academic Publishers, Dordrecht, 1998)
13. G. Hansoul, B. Teheux, *Completeness Results for Many-Valued Łukasiewicz Modal Systems and Relational Semantics* (2006). http://arxiv.org/abs/math/0612542
14. C.D. Koutras, A catalog of weak many-valued modal axioms and their corresponding frame classes. J. Appl. Non-Class. Log. **13**(1), 4772 (2003)
15. R.V. Luckheeram, R. Zhou, A.D. Verma, B. Xia, CD4+T cells: differentiation and functions. Clin. Dev. Immunol. **2012**, 12 (2012)
16. M.O. Li, Y.Y. Wan, R.A. Flavell, T cell-produced transforming growth factor-1 controls T cell tolerance and regulates Th1- and Th17-cell differentiation. Immunity **26**(5), 579–591 (2007)
17. A.M. Mironov, Fuzzy modal logics. J. Math. Sci. **128**(6), 36413483 (2005)
18. P. Ostermann, Many-valued modal propositional calculi. Zeitschrift für Mathematische Logik und Grundlagen der Mathematik **34**(4), 343354 (1988)
19. T. Porter, *Geometric Aspects of Multiagent Systems*. Electronic Notes in Theoretical Computer Science, vol. 81 (2003). http://www.elsevier.nl/locate/entcs/volume81.html
20. N. Rashevsky, *Organismic Sets* (J.M. Richards Lab, Grosse-Pointe Park, 1972)
21. R. Rosen, A relational theory of biological systems. Bull. Math. Biophys **20**, 245–260 (1958)
22. R. Rosen, The representation of biological systems from the standpoint of the theory of categories. Bull. Math. Biophys. **20**, 317–342 (1958)
23. J.D. Rutledge, A preliminary investigation of the infinitely many-valued predicate calculus. Ph.D. thesis, Cornell University, 1959
24. J.M. Smart, A.S. Kemp, Increased Th1 and Th2 allergen-induced cytokine responses in children with atopic disease. Clin. Exp. Allergy **32**, 796802 (2002)
25. N.Y. Suzuki, Kripke frame with graded accessibility and fuzzy possible world semantics. Stud. Log. **59**(2), 249269 (1997)
26. S.J. Till, J.N. Francis, K. Nouri-Aria, S.R. Durham, Mechanisms of immunotherapy. J. Allergy Clin. Immunol. **113**, 10251034 (2004)
27. H. Woodger, *The Axiomatic Method in Biology* (Cambridge University Press, Cambridge, 1937)
28. H. Yagi, T. Nomura, K. Nakamura et al., Crucial role of FOXP3 in the development and function of human $CD25^{+}CD4^{+}$ regulatory T cells. Int. Immunol. **16**(11), 1643–1656 (2004)

Potential Fields of Self Intersecting Gielis Curves for Modeling and Generalized Blending Techniques

Yohan Fougerolle, Frédéric Truchetet and Johan Gielis

Abstract The definition of Gielis curves allows for the representation of self intersecting curves. The analysis and the understanding of these representations is of major interest for the analytical representation of sectors bounded by multiple subsets of curves (or surfaces), as this occurs for instance in many natural objects. We present a construction scheme based on R-functions to build signed potential fields with guaranteed differential properties, such that their zero-set corresponds to the outer, the inner envelop, or combined subparts of the curve. Our framework is designed to allow for the definition of composed domains built upon Boolean operations between several distinct objects or some subpart of self-intersecting curves, but also provides a representation for soft blending techniques in which the traditional Boolean union and intersection become special cases of linear combinations between the objects' potential fields. Finally, by establishing a connection between R-functions and Lamé curves, we can extend the domain of the p parameter within the R_p-function from the set of the even positive numbers to the real numbers strictly greater than 1, i.e. $p \in]1, +\infty[$.

Keywords Gielis curves · R-functions · Boolean operations · Soft blending · Shape modelling

Y. Fougerolle (✉) · F. Truchetet
Le2i FRE2005, CNRS, Arts et Métiers, Univ. Bourgogne Franche-Comté, 21000 Dijon, France
e-mail: yohan.fougerolle@u-bourgogne.fr

F. Truchetet
e-mail: frederic.truchetet@u-bourgogne.fr

J. Gielis
Department of Biosciences Engineering, University of Antwerp, Groenenborgerlaan, 2020 Wilrijk-Antwerpen, Belgium
e-mail: johan.gielis@uantwerpen.be

J. Gielis et al. (eds.), *Modeling in Mathematics*,
Atlantis Transactions in Geometry 2, DOI 10.2991/978-94-6239-261-8_6

1 Introduction

The superformula, as introduced in [1], can be understood from several and complementary perspectives. For instance, it can be understood as a transform of the unit circle, i.e. a conservation law expressed in anisotropic spaces; from its initial parametric definition, the superformula appears as a polar curve which allows for the representation of many fundamental curves such as the circle, all the ellipses and superquadrics, but also numerous other natural shapes, such as fossils, crystals, flowers, etc. Under specific assumptions, the curve is closed, and therefore defines the boundary of a domain. In this context, an infinite number of potential fields can then be constructed so that the curve represents their zero-set, the interior and the exterior of the domain being characterized by potentials with opposite signs.

When the curve self intersects, the domain is split in several sectors. The numbering and the characterization of such sectors finds many applications: for instance, from a theoretical perspective, this can be applied to represent planar slices of Generalized Möbius Listing bodies GML bodies [Tavkelidze and Ricci, this volume], or as sectors of Riemann domains [2]. As a real world example, such sectors can be used to characterize the areas of growth within flowers (petals and sepals), or domains within DNA molecules. We present a summary of our work in which we proposed several techniques to build signed potential fields to characterize the interior (and exterior) of self intersecting Gielis curves. We show how the sectors due to the curve's self-intersections can be represented using nested domains and how such domains can be equipped with a signed characteristic function with guaranteed differential properties. We also propose an extended framework for blending of domains, which encompasses strict Boolean operations, that are analytically translated by R-functions, in which we now perform linear combinations between the union and the intersection of several shapes instead of traditional convex combination of their radii.

2 From Gielis Curves to Radial Distance Fields

In [1], Johan Gielis introduced the superformula, which can be seen as the parametric radius of a generalized circle, and is defined as:

$$r(\theta) = \frac{1}{\sqrt[n_1]{\left|\frac{1}{a}\cos\left(\frac{m\theta}{4}\right)\right|^{n_2} + \left|\frac{1}{b}\sin\left(\frac{m\theta}{4}\right)\right|^{n_3}}}, \tag{1}$$

with $a, b, n_i \in \mathbb{R}^+$, $n_1 \neq 0$, and $m \in \mathbb{R}^+_*$. Parameters a and b control the size, m represents the number of symmetry axes and can also be seen as the number of sectors in which the plane is folded. Coefficients n_1, n_2, and n_3 control the shape. Regular polygons (non self intersecting) as well as polygrams (self intersecting) can be approximated by setting the shape coefficients to specific values, as shown in [3].

More precisely, Maatsura proves that for any symmetry m, m not necessarily being an integer, the Gielis curve defined by $n_1 = \frac{km}{16}$ and $n_2 = n_3 = k$ converges to a regular polygon when $k \to \infty$. For the simpler case $m = 4$ and $n_2 = n_3$, the original superellipses are obtained. From this parametric formulation, an infinite number of distance-based functions can be constructed such that their zero-set corresponds to the curve. For instance, closed and non self intersecting curves can be defined as the zero-set of the following potential field:

$$F(x, y) = 1 - \frac{\left\| \overrightarrow{OP(x, y)} \right\|}{\left\| \overrightarrow{OI} \right\|} = 1 - \sqrt{\frac{x^2 + y^2}{r^2(\theta)}}, \tag{2}$$

where $I = I(O, P)$ is the intersection between the curve and the half ray originated from the center of the curve O and passing through the point $P(x, y)$, and the angle θ such that $\tan\theta = \frac{y}{x}$. When the curve is closed, the sign of such potential field can be used to define a partition of the 2D space, i.e. the set of points where $F(x, y)$ is positive corresponds to the interior of the domain bounded by the curve, the set of points where $F(x, y)$ is negative corresponds to the exterior of this domain, and the curve, that is the boundary of the domain, corresponds to the zero-set of $F(x, y)$ (Fig. 1).

Among the many possible techniques, the extension to surfaces can be obtained by the spherical product of two superpolygons, as done for superquadrics in [4]. If we note r_1 and r_2 the radii of the first and second curves, respectively, a "spherical" Gielis surface can be defined as:

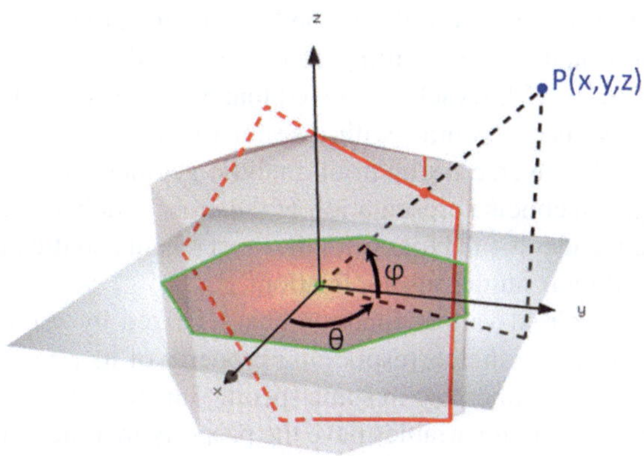

Fig. 1 Example of Gielis surface obtained as the spherical product of two Gielis curves. Using [3], the coefficients of both generating curves have been set to closely represent regular polygons. The first generating curve is the heptagon drawn in *green*, and the second one is the half pentagon, drawn in *red*. The intensity of the potential field defined in Eq. (5) is rendered on the plane $z = 0$ using discrete color coding for the positive values and smooth grey scale for the negative ones

$$\begin{pmatrix} x(\theta,\phi) \\ y(\theta,\phi) \\ z(\theta,\phi) \end{pmatrix} = \begin{pmatrix} r_1(\theta)\, r_2(\phi) \cos\theta \cos\phi \\ r_1(\theta)\, r_2(\phi) \sin\theta \cos\phi \\ r_2(\phi) \sin\phi \end{pmatrix} \tag{3}$$

From Eq. (3), Fougerolle et al. [5] proposed two potential fields for 3D non self-intersecting Gielis surfaces, that are defined in Eqs. (4) and (5), respectively.

$$F(x, y, z) = 1 - \frac{x^2+y^2+z^2 r_1^2(\theta)}{r_1^2(\theta)\, r_2^2(\phi)}, \text{ and} \tag{4}$$

$$F(x, y, z) = 1 - \frac{1}{r_2(\phi)} \sqrt{\frac{x^2+y^2+z^2}{\cos^2\phi\left(r_1^2(\theta)-1\right)+1}}, \tag{5}$$

with $\tan\theta = \frac{y}{x}$ and $\tan\phi = \frac{z r_1(\theta)\cos\theta}{x} = \frac{z r_1(\theta)\sin\theta}{y}$.

The main advantage of Eq. (5) over Eq. (4) is that the function $F(x, y, z)$ behaves as a radial distance and is therefore isotropic. These potential fields have then been used for various tasks such as solid modeling and surface reconstruction [5–7].

3 R-Functions

The theory of R-functions was introduced by Vladimir Logvinovich Rvachev in the 1960s [8]. Rvachev introduced the theory of R-functions to extend the method of Kantorovich for solving boundary value problems, i.e. to determine one representation of any arbitrary closed shape by a single analytical equation $f = 0$ or inequality $f \geq 0$ which would be satisfied only on the borders or in the interior of the object, respectively. Additionally, the resulting function would be differentiable a given number of times. In [9, 10], Rvachev showed that combining Boolean algebra—and more generally, n-valued algebra—with a set of functions which is slightly richer than polynomials, the inverse problem of analytic geometry can be solved in general. Before we give a concise presentation of the most widely used R-functions, we refer the interested reader to [11] for one of the most recent and illustrative documents on R-functions and their numerous applications.

An R-function is a real-valued function characterized by some property that is completely determined by the corresponding property of its arguments, such as for instance its sign or its membership in some specific interval. As a special case, some real-valued functions of real variables have the property that their sign is completely determined by the signs of their arguments. For example, the sign of the function $f(x_1, x_2) = x_1 x_2$ is completely determined by the sign of its arguments, as opposed to the function $f(x_1, x_2) = sin(x_1 x_2)$ where not only the sign but also the magnitude of the arguments influences the sign of the function. In simpler words, an R-function is a function that has a constant sign in each quadrant (in 2D). Equivalently, if a continuous function $f(x_1, \ldots, x_n)$ has zeros only on coordinate hyperplanes (i.e.

$f = 0$ implies that one or more $x_j = 0$; $j = 1, \ldots, n$), then f is an R-function. This key characteristic can then be used to analytically translate Boolean predicates between closed sets which are characterized by signed potential fields into one analytical equation whose sign characterizes the interior/exterior of the resulting domain, which has been extensively studied and applied in many fields related to solid and shape modelling.

We here shortly present the three most well-known branches of R-functions R_α, R_m^0, and R_p, and one of the very few n-ary R-functions known so far. More details on their differential properties and numerous examples can be found in [11, 12]. We will denote R-conjunction and R-disjunction with the same notation (symbol $\pm$), with R-conjunction obtained by considering $-$ and R-disjunction by considering $+$, respectively. R-negation is simply the opposite sign function $\overline{x} \equiv -x$. At the end of this section, we reinterpret the principles of R-functions as an analytic combination of the lengths of the sides of a triangle and then extend the formulation to allow for soft blending techniques, which encompasses strict Boolean operations represented by classical R-functions.

3.1 Systems of R-Functions and Some Properties

One of the earliest R-function proposed by Rvachev is the $R_\alpha(x_1, x_2)$ function defined as:

$$R_\alpha(x_1, x_2) : \frac{1}{1+\alpha}\left(x_1 + x_2 \pm \sqrt{x_1^2 + x_2^2 - 2\alpha x_1 x_2}\right) \tag{6}$$

where $\alpha(x_1, x_2)$ is an arbitrary symmetric function such that $-1 < \alpha(x_1, x_2) \leq 1$. Setting $\alpha(x_1, x_2) = 1$ leads to the simplest and popular R-functions: R-conjunction $min(x_1, x_2)$ and R-disjunction $max(x_1, x_2)$. However, the major drawback of these functions is that they are not differentiable along the line $x_1 = x_2$. As a consequence, two other functions, namely R_0^m and R_p are proposed to solve this lack of differentiability:

$$R_0^m(x_1, x_2) : \left(x_1 + x_2 \pm \sqrt{x_1^2 + x_2^2}\right)\left(x_1^2 + x_2^2\right)^{\frac{m}{2}}, \tag{7}$$

m is any even positive integer, and

$$R_p(x_1, x_2) : x_1 + x_2 \pm \left(x_1^p + x_2^p\right)^{\frac{1}{p}}, \tag{8}$$

p is any even positive integer.

Shapiro shows in [12] that R_0^m is m times differentiable everywhere, including the corner point $x_1 = x_2 = 0$, where partial derivatives are identically zero. Unfortunately, R_0^m is not normalized, which means that the function does not well approximate the shortest Euclidean distance to the boundary of the domain. The R_p-function achieves normalization up to $p - 1$ order, and is therefore often considered as the

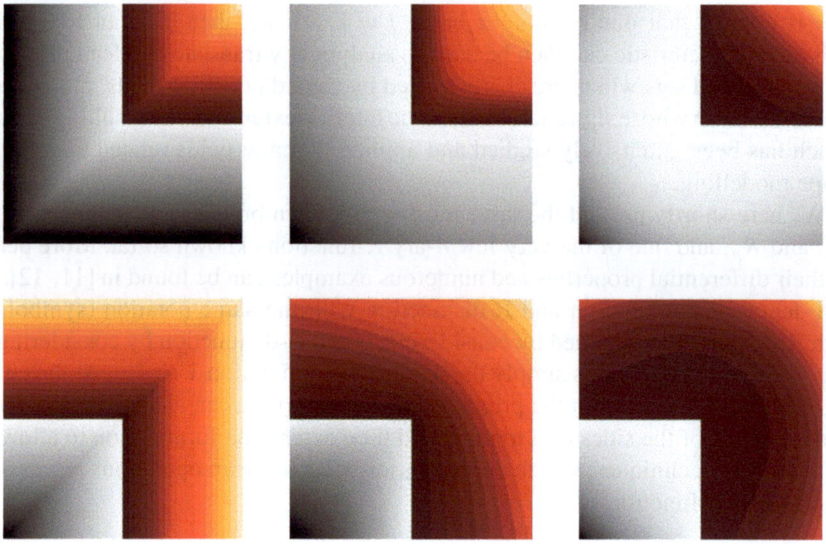

Fig. 2 Sign of $R_{\alpha=1}(x, y)$, $R_{p=2}(x, y)$, and $R^0_{m=2}(x, y)$ over a squared domain. Positive values are colored in discrete colors and negative values are in smooth grey scale. The zero set of the R-function is drawn in *green dashed lines*. *First row* represents R-conjunction and *second row* R-disjunction. All the 6 functions depicted have a constant sign in each quadrant

most convenient R-function and the most widely used in the literature. As a major result, we point out that Eq. (8) can be seen as a special case of the $R_{p>1}$-function as

$$R_{p>1}(x_1, x_2) = x_1 + x_2 \pm \sqrt[p]{|x_1|^p + |x_2|^p}, \tag{9}$$

where the term $\sqrt[p]{|x_1|^p + |x_2|^p}$ can be seen as a "Gielis-Lamé" term as defined in Eq. (1), with $m = 4, -n_1 = n_2 = n_3 = p, x_1 = \cos\left(\frac{m\theta}{4}\right)$ and $x_2 = \sin\left(\frac{m\theta}{4}\right)$. Figure 2 illustrates the intensity of these three common R-functions.

The proof that the $R_{p>1}$-function has the same sign as the $R_{p=2}$-function everywhere and is therefore an R-function can be simply obtained by inequalities between p-norms and interval bounds. We recall that according to Rvachev, the parameter p used in $R_p(x_1, x_2)$ has to be an even and positive integer. We show that such restriction on p can be alleviated, i.e. the function $R_p(x_1, x_2)$ defined by Rvachev is a special case of the R-function $R_{p>1}(x_1, x_2)$ defined as

$$R_{p>1}(x_1, x_2) = x_1 + x_2 \pm \sqrt[p]{|x_1|^p + |x_2|^p}. \tag{10}$$

It is known that for any vector x, $p > 1$ and $a \geq 0$, we have

$$\|x\|_{p+a} \leq \|x\|_p . \tag{11}$$

By first considering $0 \leq a \leq 2$, we can bind $\|x\|_{p+a}$ by:

$$\|x\|_{p+2} \leq \|x\|_{p+a} \leq \|x\|_p . \tag{12}$$

Adding the sum of all the coordinates x_i on all the terms of Eq. (12) leads to

$$\sum_i x_i + \|x\|_{p+2} \leq \sum_i x_i + \|x\|_{p+a} \leq \sum_i x_i + \|x\|_p . \tag{13}$$

When the vector x is a bi-dimensional vector, we then obtain an interval bound of $x_1 + x_2 + \|x\|_{p+a}$ as

$$x_1 + x_2 + \|x\|_{p+2} \leq x_1 + x_2 + \|x\|_{p+a} \leq x_1 + x_2 + \|x\|_p , \tag{14}$$

or more simply written

$$R_{p+2}(x_1, x_2) \leq x_1 + x_2 + \|x\|_{p+a} \leq R_p(x_1, x_2). \tag{15}$$

We have obtained an interval bound by the functions R_p and R_{p+2} which both admit identical zero sets. Consequently the function $R_{p+a}(x_1, x_2)$ also admits the same zero set, and is therefore an R-function of the same branch. By induction, we get that the function $R_{p\geq 2}(x_1, y_2)$ is an R-function. Same logic apply for the disjunction, but the order of the inequalities is swapped.

Now remains the interval $1 < p \leq 2$. Let us still consider the R-disjunction for the sake of simplicity and recall that $R_{\alpha=1}(x, y)$ corresponds to the function $max(x, y)$. It is straightforward to verify that $R_{\alpha=1}(x, y) < R_{1<p\leq 2}(x, y)$. We therefore obtain the interval bound

$$R_{\alpha=1}(x, y) \leq x + y + \|x\|_{p+a} \leq R_{p=2}(x, y), \tag{16}$$

and the same logic applies: the function is bounded by two branches of R-functions that admit an identical zero set, so it also belongs to the same branch. Similar logic can be applied for R-conjunction using the min function. Consequently, from intervals obtained in (15) and (16), we get that the function $R_{p>1}(x, y)$ is an R-function for any $p > 1$.

Because R_α, R_0^m, and R_p functions are defined to handle two arguments, these functions have to be cascaded to combine more than two arguments. Naturally, the order of composition does not alter the zero-set nor the sign of the resulting function, which is equivalent to state that the boundary of the composed object does not depend on the order used to combine the components. However, the order in which the arguments are combined strongly alters the numerical values of the resulting function everywhere else but its zero-set, as illustrated in Fig. 3. This becomes a critical issue when one seeks, for instance, to fit multiple components at the same time, because the behavior of the algorithm will strongly depend on how many functions are combined, and in which order.

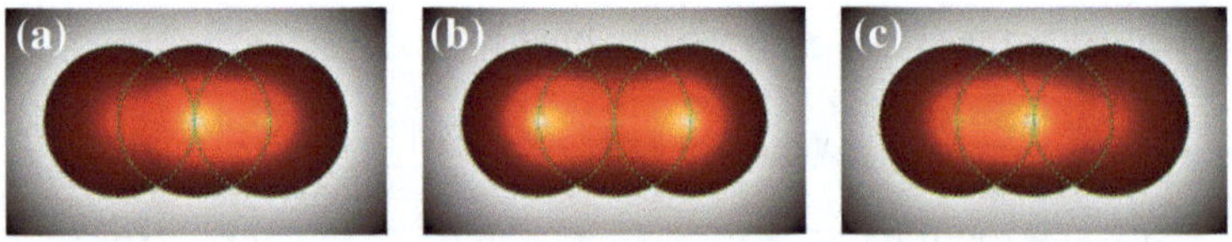

$(D1 \cup D2) \cup D3$ $(D1 \cup D3) \cup D2$ $(D2 \cup D3) \cup D1$

Fig. 3 Example of different orders of composition for 3 disks noted $D_i, i = 1, 2, 3$. Only the zero set which corresponds to the frontier between positive values in color, and negative colors in smooth grey scale remains unchanged

Besides the $min(x_1, \ldots, x_n)$ and $max(x_1, \ldots, x_n)$ functions that are commutative by definition, Rvachev also proposed some commutative R-functions to represent n-ary conjunction and disjunction. In the following definition, m is still an integer and corresponds to the parameter used to control the order of differentiability, similarly to the R_m function in Eq. (7). To our knowledge, there is very little published research that explicitly tackled the problem of the definition of commutative n-ary R-functions, and the problem remains mostly unexplored for $n > 3$.

$$\bigwedge_{i=1}^{i=n^{(m)}} x_i \equiv \sum_{i=1}^{n} (-1)^m \, x_i^m \, (x_i - |x_i|) + \prod_{i=1}^{n} x_i^m \, (x_i + |x_i|) \,. \tag{17a}$$

$$\bigvee_{i=1}^{i=n^{(m)}} x_i \equiv \sum_{i=1}^{n} x_i^m \, (x_i + |x_i|) - \prod_{i=1}^{n} (-1)^m \, x_i^m \, (|x_i| - x_i) \,. \tag{17b}$$

3.2 *From the Geometric Interpretations of Binary R-Functions to Generalized Blending*

The definitions of R_α and R_p in Eqs. (6) and (8), respectively, can also be interpreted in a geometric sense. If we think of x_1 and x_2 as the lengths of the sides of a triangle and $\alpha(x_1, x_2)$ being the cosine of the angle between these two sides, R_α and $R_{p=2}$ can be understood as a signed relationship between the lengths of the three sides of a triangle. $R_{\alpha=0}$ is often presented as an unfortunate notation from Rvachev because $R_{\alpha=0}$ and $R_{p=2}$ look similar. From the differential point of view, we already mentioned that R_α is not differentiable along all the line $x_1 = x_2$, which is not the case for R_p that is differentiable everywhere except at the origin $x_1 = x_2 = 0$. However, both functions can be interpreted as the same relationship between the lengths of the sides of a triangle that is rather expressed for the special case of the right triangle for $R_{p=2}(x_1, x_2)$ instead of the general case $R_\alpha(x_1, x_2)$. For the sake of simplicity and conciseness, we will use the function $R_{p=2}(x_1, x_2)$ but the logic presented in the following holds for any binary R-conjunction and R-disjunction.

By dividing Eq. (8) by the magnitude of its argument $\sqrt{x_1^2 + x_2^2}$, the function $R_{p=2}(x_1, x_2)$ in cartesian coordinates can be expressed in an equivalent polar form $\tilde{R}_{p=2}(\theta)$ as:

$$\tilde{R}_{p=2}(\theta) = \cos\theta + \sin\theta \pm 1 = \sqrt{2}\cos\left(\theta - \frac{\pi}{4}\right) \pm 1 \tag{18}$$

We now clearly see why such function does not depend on the magnitude of its arguments: it can be expressed as a one-dimensional function of the angle θ. To better understand why such function is actually an R-function, we now have to verify that its sign over the four quadrants of the plane is constant. From Eq. (18), it is straightforward to verify that

$$\tilde{R}_{p=2}(\theta) = \sqrt{2}\cos\left(\theta - \frac{\pi}{4}\right) - 1 \geq 0 \Leftrightarrow \theta \in \left[0, \frac{\pi}{2}\right],$$

i.e. $x \geq 0$ and $y \geq 0$, as well as

$$\tilde{R}_{p=2}(\theta) = \sqrt{2}\cos\left(\theta - \frac{\pi}{4}\right) + 1 \geq 0 \Leftrightarrow \theta \in \left[-\frac{\pi}{2}, \frac{\pi}{2}\right],$$

i.e. $x \geq 0$ or $y \geq 0$. As a consequence, the function $\tilde{R}_{p=2}$ has a constant sign over each quadrant, and therefore is an R-function.

If we now consider the ± 1 term as a special case, the Eq. (18) can be rewritten in a more general form as:

$$\tilde{R}_{p=2}(\theta, k) = \cos\theta + \sin\theta \pm k = \sqrt{2}\cos\left(\theta - \frac{\pi}{4}\right) + k, \tag{19}$$

and we can now interpret the function $\tilde{R}_{p=2}(\theta)$ as a scaled cosine function that is compared to one specific threshold $k = \pm 1$. The parameter k should belong to the interval $\left[-\sqrt{2}, \sqrt{2}\right]$, otherwise the Eq. (19) has a constant sign everywhere, which is a degenerate case of R-function and no longer allows for meaningful combination of domains characterized by signed potential fields. Figure 4 illustrates the effect of the threshold k on the sign of $R_{p=2}$.

Eventually, we can see that $k \in \left[-\sqrt{2}, \sqrt{2}\right]$ controls the angle between the line $x_1 = x_2$, i.e. $\theta = \frac{\pi}{4}$ and the half lines $\frac{\pi}{4} \pm \phi(k)$, with $\phi(k) = \arccos\left(\frac{-k}{\sqrt{2}}\right) - \frac{\pi}{4}$, as it is illustrated in Fig. 5. By tuning the parameter k, we will therefore have the ability to perform Boolean operations, as expected with $k = \pm 1$, but also many other combinations such as soft blending when $|k| < 1$, as well as we can also go beyond Boolean union and intersection for $1 < |k| < \sqrt{2}$.

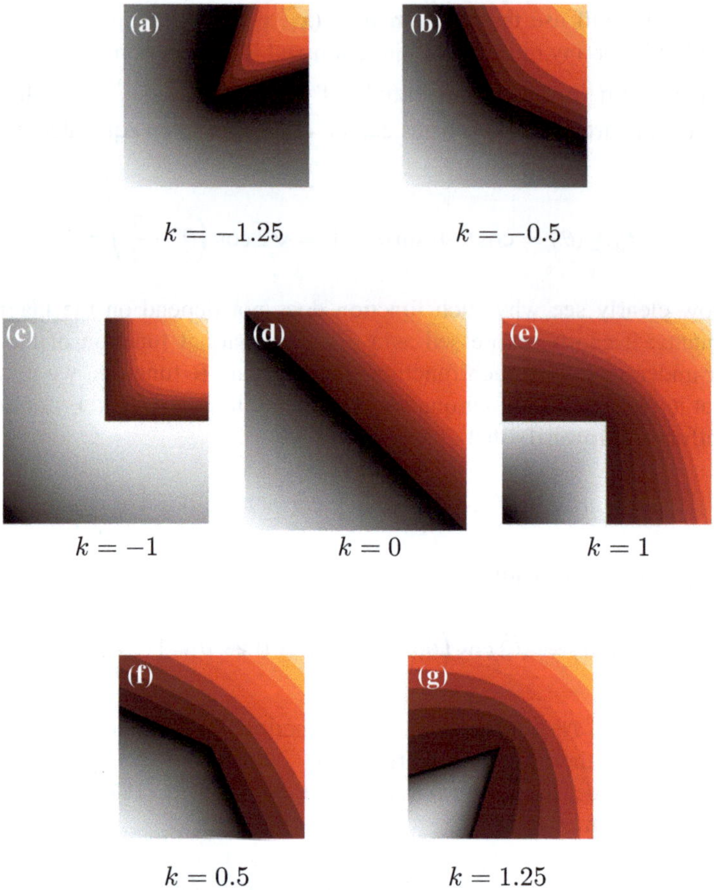

Fig. 4 Examples of function combinations for various k values in $[-\sqrt{2}, \sqrt{2}]$ using $R_{p=2}(x, y)$. The only R-functions among these examples are the functions for which their sign is constant in each quadrant and are obtained for $k = \pm 1$ (and $k = \pm\sqrt{2}$ which is a degenerate case), which yields to R-conjunction and R-disjunction as defined in Eq. (8)

As a remark, it is interesting to observe that the sum of two functions actually corresponds to the case where $k = 0$. Another way to say it is to say that the sum of two functions can be interpreted as the arithmetic mean between their conjunction and their disjunction for R_p and R_α. In a similar manner, the geometric mean between the R-conjunction and R-disjunction is equal to $\sqrt{2xy}$ for $R_{p=2}$ and $\sqrt{2xy(1-\alpha)}$ for R_α.

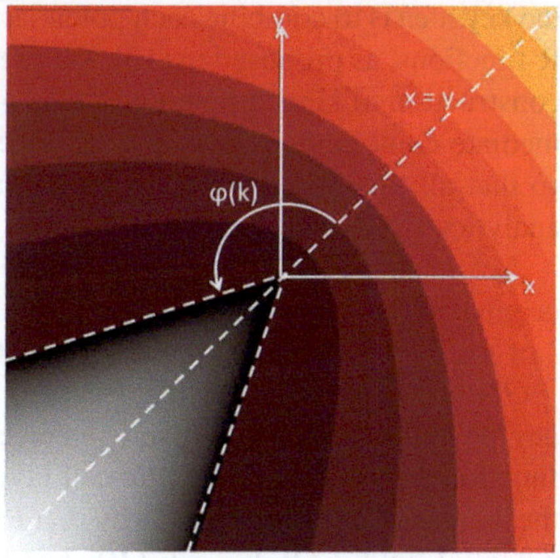

Fig. 5 Influence of parameter k on the angle $\phi(k) = \arccos\left(\frac{-k}{\sqrt{2}} - \frac{\pi}{4}\right)$ for $R_{p=2}(x, y)$

4 Potential Field Equation for Rational Gielis Curves

To introduce the rational Gielis curve (RGC), we come back to the initial formulation in Eq. (1) and now consider the symmetry parameter m as a rational number written $m = p/q$, p and q being prime relative numbers. Rational Gielis Curves are, by definition, self intersecting curves, and the maximum number of intersections between the curve and a ray initiated from its center is bounded by q, as exemplified in Fig. 6. Since we now have to handle multiple intersections, the previous definitions of the potential fields for curves and surfaces are no longer valid. To overcome this issue,

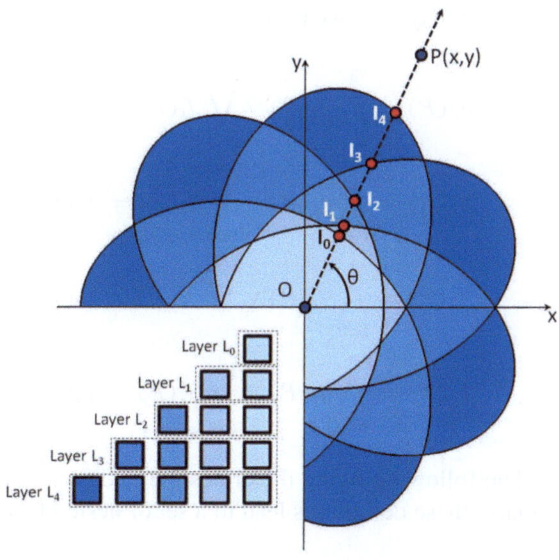

Fig. 6 Examples of self intersections with $p = 7$ and $q = 5$. The maximum number of intersections between the curve and a ray originated from the center of the curve is at most q

the key idea is to combine each potential field associated to each intersection through R-functions, as proposed in [13]. We present here a more general formulation for the construction of signed potential fields of RGC that allows to represent all the intermediate combinations between the inner "core" of the curve and its outer envelop so that all the closed subdomains that are bounded by pieces of the curve can be analytically described.

Let $I_0, I_1, \ldots, I_{q-1}$ be the q intersections between the curve and the half line originated from the center of the curve and passing by a point $P(x, y)$. The index i of each intersection I_i is set so that the distances OI_i are increasing with i. Let $f_i(P) = f_i(P(x, y)) = F(I_i, P(x, y))$ be the signed value at the point $P(x, y)$ of any potential field for non self intersecting Gielis curve, such as for instance, the function defined in Eq. (2). We call layer of index k, noted L_k, $k \in \{0, \ldots, q-1\}$, the bounded domain defined by $L_k(P) = \{P(x, y), OP(x, y) < OI_k\}$. A signed potential field $F(P, k) = F_k(P)$, $k \in \{0, \ldots, q-1\}$, with guaranteed differential properties, such that $F_k(P) > 0$ if $P \in L_k$, $F_k(P) = 0$ if P lies on the boundary of L_k, and $F_k(P) < 0$ otherwise, can be defined as follows:

$$F_k(P) = \bigwedge_{i=0}^{q-1-k} f_i(P) \vee \bigvee_{i=q-k}^{q-1} f_i(P), \tag{20}$$

with the n-ary conjunction $\bigwedge$ and disjunction $\bigvee$ being defined in Eq. (17) or being cascaded versions of the binary R-functions R_α, R_p, R_0^m, depending on the differential properties needed. For the notation to remain coherent at the extreme cases $k = 0$ and $k = q - 1$, we set $\bigwedge_0^0 X \equiv X$ and $X \vee \bigvee_{i=q}^{q-1} Y \equiv X$. In simple words, k plays the role of a cursor between the two extreme cases representing the multiple R-conjunctions and R-disjunctions of the potential fields. As an example, let consider a RGC with $q = 4$, we have:

$$F_0(P) = \bigwedge_{i=0}^{3} f_i(P) \vee \bigvee_{i=4}^{3} f_i(P) = f_0(P) \wedge f_1(P) \wedge f_2(P) \wedge f_3(P)$$

$$F_1(P) = \bigwedge_{i=0}^{2} f_i(P) \vee \bigvee_{i=3}^{3} f_i(P) = f_0(P) \wedge f_1(P) \wedge f_2(P) \vee f_3(P)$$

$$F_2(P) = \bigwedge_{i=0}^{1} f_i(P) \vee \bigvee_{i=2}^{3} f_i(P) = f_0(P) \wedge f_1(P) \vee f_2(P) \vee f_3(P)$$

$$F_3(P) = \bigwedge_{i=0}^{0} f_i(P) \vee \bigvee_{i=1}^{3} f_i(P) = f_0(P) \vee f_1(P) \vee f_2(P) \vee f_3(P)$$

The following figure illustrates the intensity of the potential fields created. As a remark, these definitions lead to a set of nested layers, i.e. $L_k \subseteq L_{k+1}$ (Fig. 7).

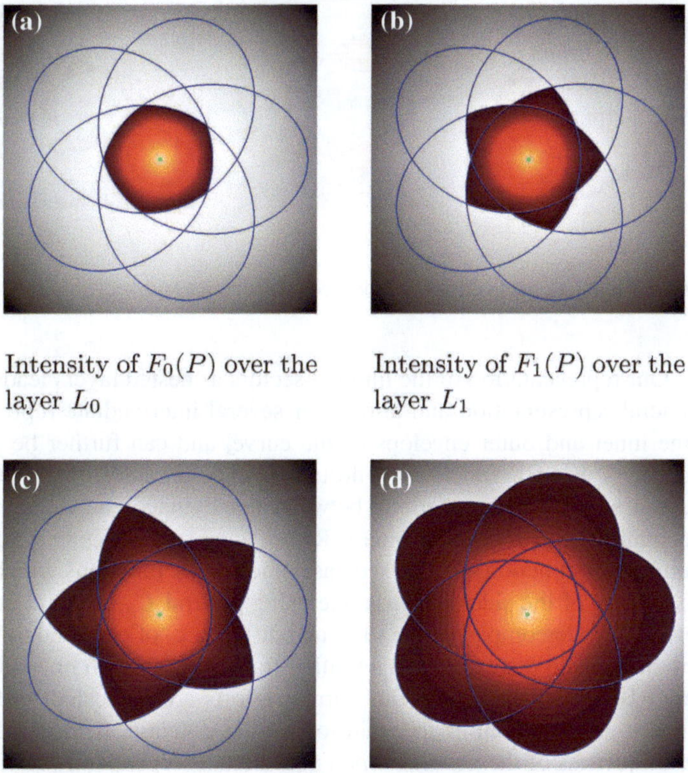

(a) Intensity of $F_0(P)$ over the layer L_0

(b) Intensity of $F_1(P)$ over the layer L_1

(c) Intensity of $F_2(P)$ over the layer L_2

(d) Intensity of $F_3(P)$ over the layer L_3

Fig. 7 Intensity of the potential fields F_k corresponding to layers L_k, $k \in \{0, 1, 2, 3\}$ for a RGC with $p = 5$ and $q = 4$. The R-function used is $R_{p=2}$

As we are now able to analytically describe each layer by a signed potential field and its boundary by pieces of parametric curves, the technique proposed in [5] can be applied to efficiently recombine the layers, mostly by performing the Boolean differences, to represent all the possible combinations of the subdomains bounded by pieces of the curve, as illustrated in Fig. 8.

5 Conclusions and Future Work

We have proposed a technique to build signed potential fields to characterize domains bounded by self intersecting Gielis curves. Thanks to the numerous properties inherited from R-functions, the generated potential fields have guaranteed differential

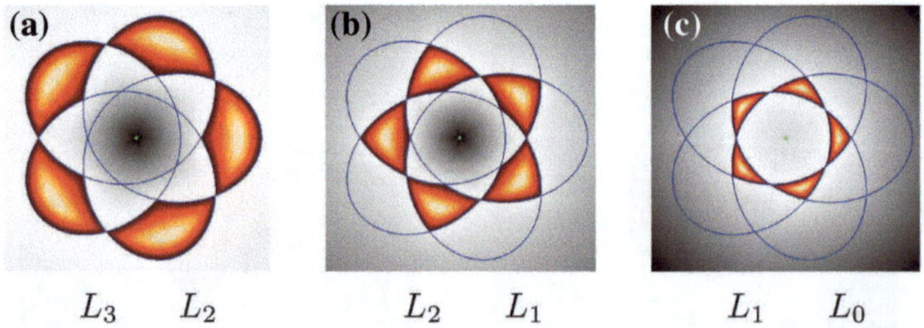

Fig. 8 Examples of the Boolean differences between consecutive layers

properties. Our representation of the internal sectors as nested layers leads to a simple and general representation and allows for several intermediate representations between the inner and outer envelops of the curve, and can further be subtracted from each others using R-functions. Throughout the summary of our studies, we have observed surprising connections between the Pythagorean theorem, the triangular inequality, most common binary R-functions and Gielis or Lamé curves. Indeed, common R-functions can be seen as some signed functions which combine the signed lengths of the sides of a triangle using general L_p-norms. Furthermore, our framework now encompasses the sum (and difference) of potential fields as well as the conjunction and disjunction in a unified framework, which now allows to go beyond strict Boolean operations. Our current efforts are dedicated to developing commutative n-ary R-functions to avoid to sort the potential fields while guaranteeing the uniqueness of the generated potential field, independently of the order of composition of the functions.

References

1. J. Gielis, A generic geometric transformation that unifies a wide range of natural and abstract shapes. Am. J. Bot. **90**(3), 333–338 (2003)
2. P. Natalini, R. Patrizi, P.E. Ricci, The Dirichlet problem for the Laplace equation in a starlike domain of a Riemann surface. Numer. Algorithms **49**(1–4), 299–313 (2008)
3. M. Matsuura, Gielis' superformula and regular polygons. J. Geom. **106**(2), 383–403 (2015)
4. A.H. Barr, Global and local deformation of solid primitives. Comput. Graph. **18**(3), 21–30 (1984)
5. Y. Fougerolle, A. Gribok, S. Foufou, F. Truchetet, M.A. Abidi, Boolean operations with implicit and parametric representation of primitives using R-Functions. IEEE Trans. Vis. Comput. Graph. **11**(5), 529–539 (2005)
6. J. Gielis, D. Caratelli, Y. Fougerolle, P.-E. Ricci, I. Tavkelidze, T. Gerats, Universal natural shapes: from unifying shape description to simple methods for shape analysis and boundary value problems. PLoS One **7**(9), 1–11 (2012)
7. Y. Fougerolle, J. Gielis, F. Truchetet, A robust evolutionary algorithm for the recovery of rational Gielis curves. Pattern Recogn. **46**(8), 2078–2091 (2013)
8. V.L. Rvachev, *Geometric Applications of Logic Algebra* (Naukova Dumka, 1967) (in Russian)
9. V.L. Rvachev, *Methods of Logic Algebra in Mathematical Physics* (Naukova Dumka, 1974) (in Russian)

10. V.L. Rvachev, *Theory of R-Functions and Some Applications* (Naukova Dumka, 1982) (in Russian)
11. V. Shapiro, Semi analytic geometry with R-functions. ACTA Numer. **16**, 239–303 (2007)
12. V. Shapiro, I. Tsukanov, Implicit functions with guaranteed differential properties, in *Symposium on Solid Modeling and Applications*, vol. 1 (1999), pp. 258–269
13. Y. Fougerolle, F. Truchetet, J. Gielis, A new potential function for self intersecting Gielis curves with rational symmetries, in *Proceedings of the International Conference on Computer Graphics Theory and Applications, GRAPP'09, Lisbon, Portugal* (2009), pp. 90–95

A Biogeometrical Model for Corolla Fusion in Asclepiad Flowers

Johan Gielis, Diego Caratelli, Yohan Fougerolle, Paolo Emilio Ricci and Tom Gerats

Abstract The molecular genetics of flower development have been studied extensively for more than two decades. Fusion of organs and the tendency to oligomery, important characteristics of flower evolution, so far have remained fairly elusive. We present a geometric model for shape and fusion in the corolla of Asclepiads. Examples demonstrate how fusion of petals creates stable centers, a prerequisite for the formation of complex pollination structures via congenital and postgenital fusion events, with the formation of de novo organs, specific to Asclepiads. The development of the corolla reduces to simple inequalities from the MATHS-BOX. The formation of stable centers and of bell and tubular shapes in flowers are immediate and logical consequences of the shape. Our model shows that any study on flowers, especially in evo-devo perspective should be performed within the wider framework of polymery and oligomery and of fusion and synorganization.

J. Gielis (✉)
Department of Biosciences Engineering, University of Antwerp,
Groenenborgerlaan 171, 2020 Antwerpen, Belgium
e-mail: Johan.gielis@uantwerpen.be

D. Caratelli
The Antenna Company Nederland B.V., High Tech Campus 41, Eindhoven,
The Netherlands

D. Caratelli
Tomsk Polytechnic University, 84/3 Sovetskaya Street, 634050 Tomsk, Russia

Y. Fougerolle
LE2I FRE2005, CNRS, Arts et Métiers, Université Bourgogne Franche-Comté,
21000 Dijon, France

P.E. Ricci
International Telematic University UniNettuno, Corso Vittorio Emanuele II, 39,
00186 Roma, Italia

T. Gerats
Radboud University Nijmegen, Section Plant Genetics, Institute for Wetland
and Water Research, Faculty of Science, Nijmegen, The Netherlands

J. Gielis et al. (eds.), *Modeling in Mathematics*,
Atlantis Transactions in Geometry 2, DOI 10.2991/978-94-6239-261-8_7

1 Introduction

Flowers enabled the evolvement of insect-mediated pollination, a driving force in the rapid evolution and radiation of both angiophytes and pollinators. Our knowledge of flowers, their diversity, development and rapid evolution is still very limited. Despite the general applicability of the basics of the *(A)BC* model [1–5] we are only at the very beginning of understanding development and evolution of flowers [6]. Improved understanding of the flower as a determining factor in the evolution of angiosperms, should involve the study of specific trends in the evolution of floral morphology [7]. Like the *(A)BC* model, such trends are an abstraction of nature, but they have the advantage to result from observations spanning the whole group of flowering plants and their ancestors, not only model plants.

One major trend in the evolution of the flower is the reduction of polymery to oligomery. Polymery is associated with spiral phyllotaxy, an undifferentiated perianth, the presence of staminodes, complex floral vasculature, and typically trimery or dimery as the basic merosity of the flower [8]. Oligomery is associated with whorled phyllotaxy and a basic condition of diplostemony [9]. A second major trend is synorganization, in which two or more specific organs or structural elements undergo fusion, within and between whorls, thus enabling the conception of new functional entities [7, 10]. A whorled arrangement, radial symmetry, and a small, fixed number of floral organs (oligomery) are the major prerequisites for the evolution of complex synorganizations [11, 12]. Synorganized structures are more stable and enable the formation of larger flowers, opening up new ways to experiment with shape and function, such as the formation of three-dimensional corollas with increased petal surface, enabling a wide spectrum of pollination syndromes [12].

Inevitably the genetics of fusion will be unraveled but at present, to our knowledge, the only molecularly defined mutant that exhibits lack of fusion of both carpels and petals is the *Petunia maewest* mutant [13, 14]. In *Arabidopsis* fusion of sepals and petals is observed only in a specific mutant [15]. For fusion to occur, growing organs must come into contact early during development and both size and spatial separation of individual organs are important. The prevalence of fusion in the perianth in dicots is associated with pentamery as the basic merosity, whereas in tetramerous flowers of *Arabidopsis* and in *P. maewest* mutants, the petals are spaced too far apart to initiate fusion [16].

Beyond simple fusion of petals, synorganisation provides for an enormous potential for the diversification of flowers [7, 10, 12]. The best example is the development of a highly complex pollination apparatus in sapromiophylous flowers with putrid or carrionlike odours (Asclepiadaceae and Rafflesiaceae) [17, 18]. With the presence of fascinating colors in connection with special structures fostering deceit, trapping and attachment of pollinia, the Asclepiadaceae are also known as the orchids among the dicotelydons [10], with flower sizes ranging from a few millimeters to forty centimeters. Flowers of *Rafflesia* can measure up to 1 m.

In Stapeliads (Asclepiadoideae) it is the corolla that accounts for this size variation, not the inner organs. The corolla originates from fusion of petals and forms a very stable structure necessary for the development of the highly complex pollination apparatus located in the center. This apparatus results from a very complicated and highly coordinated sequence of fusion events, involving congenital and postgenital fusion of petals partly released at anthesis, the congenital union of stamens followed by fusion of stamens with petals into a corona, the postgenital fusion of carpels and the formation of a style head, followed by fusion of stamens with style head, forming a gynostegium. From these, guide rails and pollinaria are formed, so that the pollinarium (two pollinia and a de novo organ, the translator, formed exclusively from secretion) are partly gynoecial, partly androecial [12].

In order to coordinate such complex developmental processes, stability is a key prerequisite and within the Stapeliads (as well as the other +/−250 genera and 3400 species of the Asclepiadaceae) the mutual position of the floral parts is very precisely fixed. Deviations from pentamery or changes in the pollination apparatus are extremely rare [12, 18]. Such stability is not found in any other Angiosperm family and has led to the formation of unique organs, not present in any other family [12, 18]. Hence, Stapeliads provide for an excellent model in the *physical* sense (simple, stable and reproducible) and in the *biological* sense (stable and reproducible, yet highly complex). Here we show that fusion in the corolla of Stapeliads can be described by a *geometrical* model based on two opposing forces, in a single transformation (Eq. 1 [19]) referred to as Gielis transformation in literature and that the occurrence of a stable center is a direct consequence of fusion and synorganization of the petals.

2 Gielis Flower Domains and Associated Laplacian

Gielis curves (Eq. 1 [19]) can be understood as transformations, where a constraining function ***CF*** acts on developable functions ***DF*** $= f(\vartheta)$ in Eq. 1.

$$\varrho(\vartheta; f(\vartheta), m, A, B, n_1, n_2, n_3) = f(\vartheta) \cdot \left[\left| \frac{1}{A} \cos\left(\frac{m}{4}\vartheta\right) \right|^{n_2} \pm \left| \frac{1}{B} \sin\left(\frac{m}{4}\vartheta\right) \right|^{n_3} \right]^{-\frac{1}{n_1}} \quad (1)$$

These transformations impose a constraining geometry in which the developable function DF (circle, spiral....) can develop, in an anisotropic way with preferred directions and distances (for $CF \neq 1$). If $CF = 1$ a circle results.

The coordinate functions of Gielis curves are obtained when $f(\vartheta)$ are *cosine* or *sine* functions. At the same time such functions are generalized trigonometric functions allowing for widening the scope of Grandi-flowers or rose-curves [19]. When using Grandi-curves equation 1 is slightly modified with the factor $m/2$ because of the absolute values necessary for the Grandi-curve. m defines the number of petals, n_4 increases petal size, $n_{1,2,3}$ determine the shape of the circumscribing (constraining) polygon.

$$\rho = \frac{DF}{(C_1 + C_2)^{\frac{1}{n_1}}} = DF \cdot CF \tag{2}$$

$$DF = \left|cos\frac{m\vartheta}{2}\right|^{n_4}; \quad CF = (C_1 + C_2)^{-\frac{1}{n_1}}; \quad C_1 = \left|cos\frac{m\vartheta}{2}\right|^{n_2}; \quad C_2 = \left|sin\frac{m\vartheta}{2}\right|^{n_3}$$

In three dimensions we define surfaces and volumes by two perpendicular 2D supershapes, one defining a supershape $\varrho_1(\vartheta)$ in the *XY* plane, the other one defines a supershape $\varrho_2(\varphi)$ in the *XZ* plane.

$$\begin{cases} x = \varrho_1(\vartheta)\cos\vartheta \cdot \varrho_2(\varphi)\cos\varphi \\ y = \varrho_1(\vartheta)\sin\vartheta \cdot \varrho_2(\varphi)\cos\varphi \\ z = \varrho_2(\varphi)\sin\varphi \end{cases} \tag{3}$$

Using the Gielis formula to describe domains, it has become possible to derive analytic solutions of boundary value problems in terms of Fourier series ([20–24]; this volume). Here we present the Fourier–Bessel solution for the Helmholtz equation in flower corolla domains whereby the corolla domain is considered as a plate. The plate vibration problem can be solved using two Helmholtz equations, using the following expression of the Laplacian.

Let us consider in the Euclidean plane the classical polar coordinate system

$$\begin{cases} x = \rho\cos\vartheta \\ y = \rho\sin\vartheta \end{cases}$$

and the starlike domain D, whose boundary ∂D is described by the equation $\rho = R(\vartheta)$, where $R(\vartheta)$ is a C^2 function in $[0, 2\pi]$, such that $m_R = min_{\vartheta\in[0,2\pi]}R(\vartheta) > 0$, and $M_R = max_{\vartheta\in[0,2\pi]}R(\vartheta) < 1$. Therefore, upon introducing the stretched radius: $r = \frac{R(\vartheta)}{\rho}$ it is straightforward to see that D satisfies $0 \leq \vartheta \leq 2\pi$, and $0 \leq r \leq 1$.

Let us consider a $C^2(D)$ function of $v(x, y) = v(\rho\cos\vartheta, \rho\sin\vartheta) = u(\rho, \vartheta)$ and the relevant Laplacian operator

$$\Delta u = \frac{\partial^2 u}{\partial\rho^2} + \frac{1}{\rho}\frac{\partial u}{\partial\varrho} + \frac{1}{\rho^2}\frac{\partial^2 u}{\partial\vartheta^2}$$

Upon setting $u(r, \vartheta) = u(rR(\vartheta), \vartheta)$, we readily find:

$$\frac{\partial u}{\partial\rho} = \frac{1}{R}\frac{\partial U}{\partial r}, \quad \frac{\partial^2 u}{\partial\rho^2} = \frac{1}{R^2}\frac{\partial^2 U}{\partial r^2}, \quad \frac{\partial u}{\partial\vartheta} = -r\frac{R'}{R}\frac{\partial U}{\partial r} + \frac{\partial U}{\partial\vartheta},$$

$$\frac{\partial^2 u}{\partial\vartheta^2} = -\frac{R'^2 - RR''}{R^2}\frac{\partial U}{\partial r} + r^2\frac{R'^2\partial^2 U}{R^2\partial r^2} - 2r\frac{R'}{R}\frac{\partial^2 U}{\partial r\partial\vartheta} + \frac{\partial^2 U}{\partial\vartheta^2}$$

Hence, substituting, we obtain

$$\Delta u = \frac{\partial^2 u}{\partial \rho^2} + \frac{1}{\rho}\frac{\partial u}{\partial \varrho} + \frac{1}{\rho^2}\frac{\partial^2 u}{\partial \vartheta^2} = \tag{4}$$

$$\frac{1}{R^2}\left[1+\frac{R'^2}{R^2}\right]\frac{\partial^2 U}{\partial r^2} + \frac{1}{rR^2}\left[1+\frac{2R'^2 - RR''}{R^2}\right]\frac{\partial U}{\partial r} - \frac{2R'}{rR^3}\frac{\partial^2 U}{\partial r \partial \vartheta} + \frac{1}{r^2R^2}\frac{\partial^2 U}{\partial \vartheta^2}$$

For $r = \rho$, $R(\vartheta) \equiv 1$ we recover the Laplacian in usual polar co-ordinates.

3 Leaf Shapes and Phyllotaxy

Gielis transformations introduce global anisotropy and can transform a planar function $f(\vartheta)$, for example a circle (for $(f\vartheta)$ = constant) into squares, starfish or indeed any regular, concave or convex polygon. Alternatively, when the transformation is constant, $f(\vartheta)$ develops in the isotropic, Euclidean case and assumes its well-known shape, the Euclidean circle. Various leaf types (linear, oblong, rhombic, lanceolate, ovate, elliptic, cordate, hastate, cuneate, spatulate, saggitate, deltoid, reniform), can be described by Eq. 1 [25, 26]. Many leaf shapes can be considered as transformations from a basic cardioid $r = 1 + \cos\vartheta$. A cardioid is in fact a combination of shapes, the sum of a unit circle with a second unit circle with its center on its boundary. Using Eq. 1 as a transformed cardioid various shapes in leaves are easily obtained (Fig. 1). This allows for modeling development of leaves as a concentric, yet anisotropic development. In Fig. 1 the petiole is attached to the leaf blade in the origin of the cardioid.

In the 3D case, all shapes defined by Eq. 1 can serve as profiles of cross sections, stacked along a central axis. In this way it is straightforward to consider plant stems as generalized cylinders, with changing cross-profiles along a central axis. The axis does not need to be a straight line. It can be curved, as in canal surfaces, or closed as in Generalized Möbius-Listing surfaces and bodies ([28]; this volume). Along such generalized cylinders or GML surfaces and bodies definite phyllotaxies can evolve or be defined [25, 29].

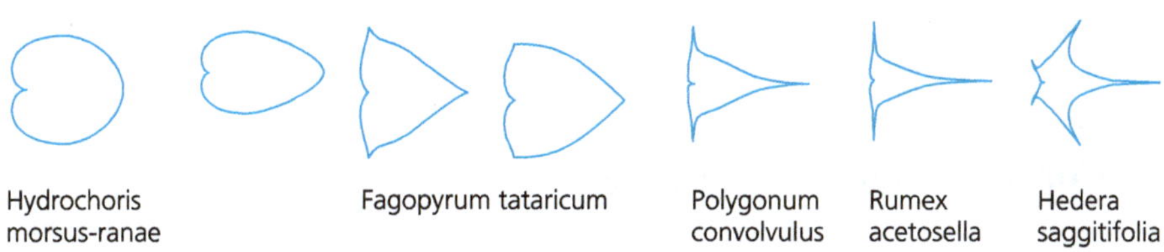

Fig. 1 Leaves of various plants, as variations on a basic theme [27]

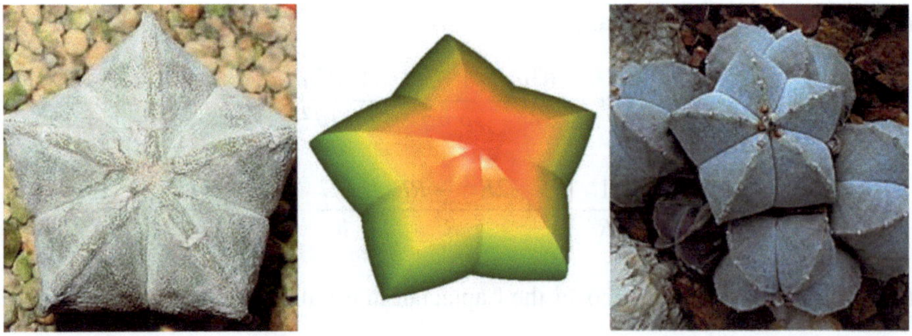

Fig. 2 *Left* and *right* Astrophytum, *Center* a supershape based on two sections, with vertical section a circle, and horizontal section a 5/2 polygram

In wild roses (*Rosa canina* or dog rose) positioning of leaves along the cylindrical stem is 5 leaves in two rotations, leaves spaced 144° apart. This corresponds to 2/5 as consecutive Fibonacci numbers, but the planar projection of this structure is nothing but a pentagram defined by Eq. 1 with m = 5/2 (five angles in two rotations, [19]). This spiral phyllotaxy is then condensed into a planar structure in the sepals, whereby the spiral phyllotaxy is still observed in the initiation of sepals during development. In the next whorl, the initiation of the five petals is no longer spiral but whorled ($m = 5/1$; five petals in one rotation, petals spaced 72° apart). In general the projection of phyllotactic patterns in a plane can be represented by Eq. 1, where the symmetry parameter *m* is the inverse of the common ratio of Fibonacci numbers. Since *m* is a real number also small differences caused by genetic factors or developmental deviations, can be modeled. Indeed, the cycle does not to close exactly at 360° or $k \cdot 2\pi$ with *k* an integer number, as observed in spiromonostichy and spirodistichy or irregular phyllotactic patterns.

This provides also a geometrical basis for understanding a range of characteristics of cacti. In the simplest columnar cacti, the cross sections are shapes with integer symmetry (*m* is integer), but spiral phyllotaxy can be studied in the same way. *Astrophytum* (Fig. 2) is one example where spiral phyllotaxy in a plane is petrified into a sphere-like plant. Its three dimensional shape is described by two perpendicular sections with $m = 5/2$ and exponents $n = 1$ in horizontal section ρ_1, and a circle ρ_2, in the perpendicular, vertical section, whereby ρ_1 and ρ_2 are defined by Eq. 3. This shows that along the proximal-distal axis the horizontal plane needs not to remain constant.

4 Corolla Shapes in Asclepiad Flowers

The shape of the corolla of Asclepiad flowers can also be described by Gielis curves and transformations. These transformations impose a global anisotropy and preferred direction on $f(\vartheta)$. In other words, they 'constrain' the growth of a function $f(\vartheta)$ that wants to 'expand' or 'develop'. We call these two opposing functions the Constraining Function *CF* and the Developing Function *DF*, respectively.

When *DF* is a cosine function, rose or Grandi curves result, used to model certain flowers. When constrained by Gielis transformations *GT* they easily model a much wider variety of flowers. Indeed, by multiplying a rose curve *DF* with a Constraining Function *CF*, a wide range of flowers can efficiently be described. The petals in such flowers are free standing, a condition known as choripetaly. Choripetaly is observed for example in Solanaceae, strawberry and *Geranium* (upper row of Fig. 3). Zygomorphy can be obtained when rose curves are inscribed in asymmetric Gielis curves (for example monogon or egg-like shapes for $m = 1$). In Fig. 3 and Table 1, the flowers left and center left differ only in *DF* (petal size n_4), not in *CF*. The flowers center right and right differ in the circumscribing *CF* ($n_{2,3}$) only. The superpolygon center right is inscribed in the base circle, the other one circumscribes the basic circle.

In sympetalous flowers petals are fused to at least some extent and the operation is not multiplication, but addition. Sympetalous flowers or Asclepiad corollas are obtained by weighted addition of the two functions *CF* and *DF*, defined by a weight parameter α (Eq. 5), with $0 \leq \alpha \leq 1$ (here C_1 and C_2 have $m/4$). While technically Eq. 5 defines the coordinate system for the supershape defined in Eq. 1, here we use

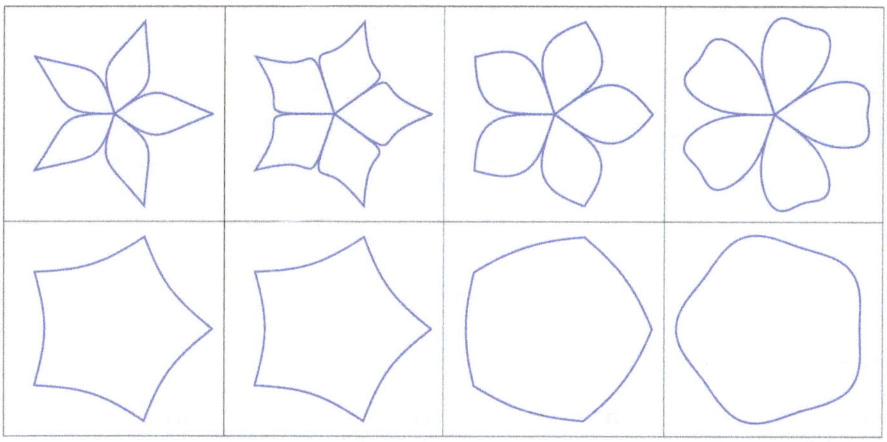

Fig. 3 Choripetalous flowers with the corresponding constraining superpolygons. Values of the various parameters in Table 1

Table 1 Values of parameters for symmetry (m) and the exponents (n_i) in Fig. 3. The *upper row* is given by Eq. 2, the corresponding *CF*'s in the *lower row* by Eq. 1 (with $DF = 1$)

m	5	5	5	5
n_1	*1*	*1*	*3*	*3*
$n_{2,3}$	*1*	*1*	*1*	*3*
n_4	2	*5*	*3*	*3*

the polar coordinates system, whereby *DF* is the developable function and *CF* the constraining function. The interplay between both (Eq. 5) defines the shape (Fig. 4).

$$R(\vartheta) = (1-\alpha)\left|cos\frac{m\vartheta}{2}\right|^{n_4} + \alpha\left[\left|\cos\left(\frac{m\vartheta}{4}\right)\right|^{n_2} \pm \left|\sin\left(\frac{m\vartheta}{4}\right)\right|^{n_3}\right]^{-\frac{1}{n_1}}$$
$$= (1-\alpha)\cdot DF + \alpha \cdot CF \tag{5}$$

The influence of *DF* and *CF* (Fig. 5) is dependent with a weight parameter α. When α is close to one, petals will almost completely be fused, like in *Caralluma frerei* (Fig. 5a,d). In *Huernia recondita* (Fig. 5c) both functions contribute about half (i.e. $\alpha = 0.5$), the arithmetic mean AM of *CF* and *DF*.

Increasing fusion has a direct effect on area, perimeter, polar moment of inertia and area use efficiency AUE, which is defined as the ratio between the area of the *DF* and the area of the underlying constraining geometry *CF*. In the case of choripetaly in Fig. 3 this can be observed clearly. In the case of sympetaly the AUE for $\alpha = 1$ is

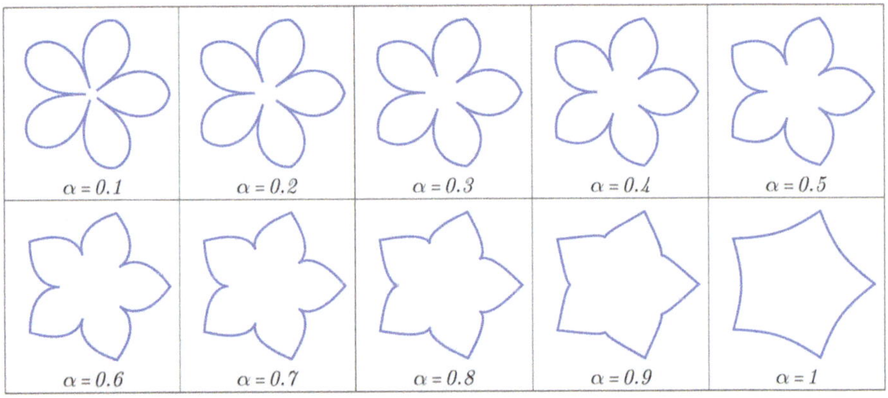

Fig. 4 Increasing fusion of petals

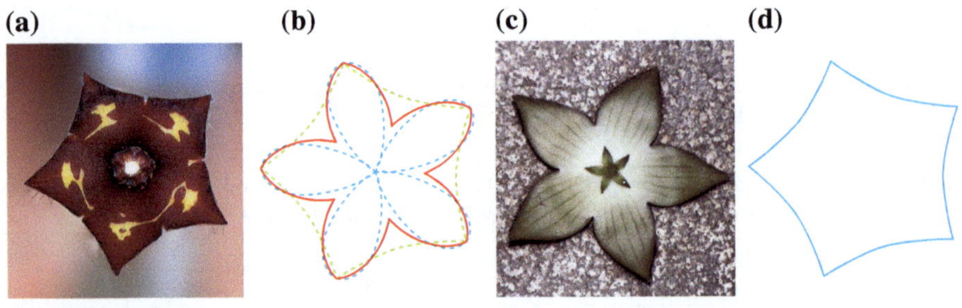

Fig. 5 **a** *Caralluma frerei*. **b** *CF* (*green*), *DF* (*blue*) and fusion (*red*) for $\alpha = 0.5$. **c** *Huernia recondita*. **d** *C. frerei* with $\alpha = 1$

higher than for lower values of α (Fig. 5). In Fig. 5d the corolla of Caralluma has a higher AUE than the corolla of Huernia.

5 A Three-Element System for Fusion in Corolla

Using the two basic operations from the MATHS-Box between *DF* and *CF*, the shape of any flower can be described by three strategies (*X*, *Y* and *Z*). The first strategy, *X*, defines *DF* (number and size of petals) and *CF* (constraining polygon). *Y* and *Z* involve multiplication and addition of *CF* and *DF* respectively, allowing for the description of choripetalous and sympetalous flowers (Figs. 3, 4 and 5). Combining *X*, *Y* and *Z* in a three-component system, whereby *X* is present in all combinations completes our model (Table 2). *Thunbergia alata* (Fig. 6) is one example of a combination of *X*, *Y* and *Z* strategies.

Table 2 State table for *(X)YZ* and *(A)BC* model with four possibilities. The *A* and *X* states are always on (1)

	X	*Y*	*Z*	Strategy	State
3*0	0	0	0	0	
2*0	1	0	0	X	
1*0	1	1	0	*X*°*Y*	Choripetaly
	1	0	1	*X*°*Z*	Sympetaly
0*0	1	1	1	*X*°*Y*°*Z*	Most complete
	A	*B*	*C*		State
3*0	0	0	0	0	No flower
2*0	1	0	0	*A*-Only	Only sepals
1*0	1	1	0	*A*+*B*(Class *C*)	S-P-P-S
	1	0	1	*A*+*C*(Class *B*)	S-S-C-C
0*0	1	1	1	*A*+*B*+*C*	Most complete

Fig. 6 Black-eyed Susan (*Thunbergia alata*)

As a three-element system, X, Y and Z can be applied like the well-known *(A)BC* model for molecular aspects of flower development. The *ABC* model was initially defined based on the analysis of morphological mutants in only two species. Though readily applicable in its basics, nature should be expected to have been (and remain) flexible with regard to the genetic set up of developmental processes. Indeed, the rather diverged control of the *A* function [4, 5, 30] has resulted in a revision of the *(A)BC* model. The *A*-function establishes a floral context by its involvement in the control of floral meristem identity; it provides the requirements necessary to enable the *B*- and *C*-function genes to exert their control over floral organ identity [5].

In a similar way, the (*X*) strategy controls the individual functions (global anisotropy *CF*, and petal size and number *DF*) and fulfills the requirements for the ultimate outcome of the *Y* and *Z* strategies, which are multiplication and addition of these functions respectively. *X* differs from *Y & Z,* as *A* differs from *B & C* (Table 2). Under the conditions that one element always is switched on, a three element system yields only four instead of eight developmental variants, which is one of the reasons why the original *ABC* model has to be substituted for by the *(A)BC model* [1].

6 Fourier–Bessel Solution for the Helmholtz Equation

This description allows for developing general methods to study shape and changes of shape, when certain forms of stress or tension are applied, using partial differential equations using the Laplacian. Using the transformations to stretch the unit circle also allows to define Laplacian in these stretched coordinates and to obtain analytic solutions for various boundary value problems using Fourier's original projection method.

The Helmholtz equation is used in various problems in mathematical physics involving space and time, such as in wave equation and the study of membranes or plate vibrations [31]. The fleshy and thick corollas of Asclepiads may be considered as semi-rigid plates, rather than as membranes, but since the Helmholtz equation is involved in both, we focus on deriving the general, analytical solution for the Helmholtz equation. The displacement w of points on a vibrating membrane is given by the classical differential equation

$$D\nabla^4 w + \frac{\varrho \partial^2 w}{\partial t^2} = 0$$

with flexural rigidity $D = \frac{Eh^3}{12(1-\nu^2)}$ (E is Young's modulus, h thickness of the plate, ν Poisson's ratio and ρ mass density). For free vibrations we have $w = Wcos(\omega t)$ with W a function of position coordinates. This gives: $(\nabla^4 - k^4)W = 0$, with $k^4 = \frac{\rho\omega^2}{D}$, or $(\nabla^2 + k^2) \cdot (\nabla^2 - k^2)W = 0$. The complete solution can thus be obtained by superimposing solutions to the equations:

$$\begin{cases} \nabla^2 W_1 + k^2 W_1 = 0 \\ \nabla^2 W_2 - k^2 W_2 = 0 \end{cases}$$

Let us consider the interior Dirichlet problem for the Helmholtz equation in a starlike domain D, whose boundary is described by the polar equation $\rho = R(\vartheta)$,

$$\begin{cases} \Delta v(x,y) + k^2 v(x,y) = 0 & (x,y) \in D \\ v(x,y) = f(x,y) & (x,y) \in \partial D \end{cases}$$

with k denoting the propagation constant.

The following theorem can be proved:

Theorem 1 – Let

$$f(R(\vartheta)\cos\vartheta, R(\vartheta)\sin\vartheta) = F(\vartheta) = \sum_{m=0}^{+\infty} (a_m \cos m\vartheta + \beta_m \sin m\vartheta)$$

where

$$\begin{Bmatrix} \alpha_m \\ \beta_m \end{Bmatrix} = \frac{\varepsilon_m}{2\pi} \int_0^{2\pi} F(\vartheta) \begin{Bmatrix} \cos m\vartheta \\ \sin m\vartheta \end{Bmatrix} d\vartheta$$

ε_m being the usual Neumann's symbol. Then, the interior boundary-value problem for the Helmholtz equation admits a classical solution. $v(x,y) \in C^2(D)$ such that the following Fourier–Bessel expansion holds

$$v(rR(\vartheta)cos\vartheta, rR(\vartheta)sin\vartheta) = U(r,\vartheta) = \sum_{m=0}^{+\infty} J_m(krR(\vartheta))(A_m \cos m\vartheta + B_m \sin m\vartheta) \quad (6)$$

with $J_m(.)$ denoting, as usual, the Bessel function of the first kind and order m. The coefficients A_m, B_m can be determined by solving the infinite linear system

$$\sum_{m=0}^{+\infty} \begin{bmatrix} X^{+}{}_{n,m} & Y^{+}{}_{n,m} \\ X^{-}{}_{n,m} & Y^{-}{}_{n,m} \end{bmatrix} = \begin{bmatrix} \alpha_m \\ \beta_m \end{bmatrix},$$

where

$$\begin{cases} X_{n,m}^{(\pm)} = \frac{\varepsilon_n}{2\pi} \int_0^{2\pi} J_m(kR(\vartheta) \cos m\vartheta \begin{Bmatrix} \cos n\vartheta \\ \sin n\vartheta \end{Bmatrix} d\vartheta \\ Y_{n,m}^{(\pm)} = \frac{\varepsilon_n}{2\pi} \int_0^{2\pi} J_m(kR(\vartheta) \sin m\vartheta \begin{Bmatrix} \cos n\vartheta \\ \sin n\vartheta \end{Bmatrix} d\vartheta \end{cases}$$

with $m, n \in \mathbb{N}_0$.

Proof: In the stretched co-ordinate system (r, ϑ) the domain D is transformed in the unit circle. Therefore, the usual eigenfunction method and separation of variables (with respect r and ϑ) can be conveniently used to solve Eq. 6. As a consequence, elementary solutions of the problems can be searched in the form:

$$u(\rho,\vartheta) = U\left(\frac{\rho}{R(\vartheta)},\vartheta\right) = P(\rho)\Theta(\vartheta)$$

By substituting into the Helmholtz equation, one can easily find out that the functions $P(\rho)\Theta(\vartheta)$ must satisfy the ordinary differential equations:

$$\frac{d^2\Theta(\vartheta)}{d\vartheta^2} + \mu^2\Theta(\vartheta) = 0,$$

$$\rho^2\frac{d^2P(\rho)}{d\rho^2} + \rho\frac{dP(\rho)}{d\rho} + \left[(k\rho)^2 - \mu^2\right]P(\rho) = 0, \text{ respectively.}$$

The parameter μ is a separation constant whose choice is governed by the physical requirement that at any fixed point in the plane the scalar field $u(\rho,\vartheta)$ must be single-valued. By setting $\mu = m \in \mathbb{N}_0$ we find:

$$\Theta(\vartheta) = a_m \cos m\vartheta + b_m \sin m\vartheta$$

when $a_m, b_m \in \mathbb{R}$ denote arbitrary constants. The radial function $P(.)$ satisfying the ODE can be readily expressed as follows:

$$P(\rho) = c_m J_m(k\rho) + d_m Y_m(k\rho) \quad (c_m, d_m \in \mathbb{R})$$

As usual, we have to assume $d_m = 0$ for the boundedness of the solution.

Therefore, the general solution of the interior Dirichlet problem can be expressed in the form:

$$u(\rho,\vartheta) = \sum_{m=0}^{\infty} J_m(k\rho)(A_m \cos m\vartheta + B_m \sin m\vartheta)$$

Enforcing the boundary conditions yields:

$$F(\vartheta) = U(1,\vartheta) = u(R(\vartheta),\vartheta) = \sum_{m=0}^{\infty} J_m(kR(\vartheta))(A_m \cos m\vartheta + B_m \sin m\vartheta)$$

and finally, using Fourier' projection method, the system of linear equations easily follows. The developed method can be applied to the solution of differential problems in a two-dimensional flower shaped domain D, whose boundary is described by the fusion-like Gielis curve:

$$\rho = R(\vartheta) = (1-\alpha)\left|\cos\frac{m\vartheta}{2}\right|^{n_4} + \alpha\left[\left|\cos\left(\frac{m\vartheta}{4}\right)\right|^{n_2} \pm \left|\sin\left(\frac{m\vartheta}{4}\right)\right|^{n_3}\right]^{-\frac{1}{n_1}}$$

As an example, by using the Theorem the interior Dirichlet problem for the Helmholtz equation has been solved upon setting $m = 5$, exponent $n_i = 1$ and $a = 2/5$. Furthermore the boundary values have been assumed to be given by the radially symmetric function

$$f(x, y) = (x^2 + y^2)^\tau exp[(x^2 + y^2)/\sigma^2], \text{ with } \tau = 3 \text{ and } \sigma = \frac{1}{2}$$

In particular the numerical accuracy has been tested and assessed in terms of the following boundary error.

$$e_N = \frac{\| U_N(1, \vartheta) - F(\vartheta) \|}{\|F(\vartheta) \|},$$

with $\|.\|$ denoting the usual L^2 norm, U_N the partial sum of order N relevant to the Fourier-like series expansion representing the solutions of the Dirichlet problem for the Helmholtz equation, and F the function describing the boundary values (Figs. 7, 8 and 9).

7 Leaves and Corollas

Equation 1 is a generalization of Lamé curves and conic sections more generally, keeping the structure of the Pythagorean Theorem with a priori separation of variables in the plane. Gielis Transformations are increasingly used in geometry [32], in applied mathematics [20–24] and in technology [33–35]. In biology they are used e.g. to model bamboo leaves [36, 37], petioles in *Philodendron* and rhubarb [38, 39], tree rings in conifers [40], and bamboo culms, rhizomes and meristems [41]. In disguise, supershapes and Lamé curves are characteristic of natural networks, whereby power

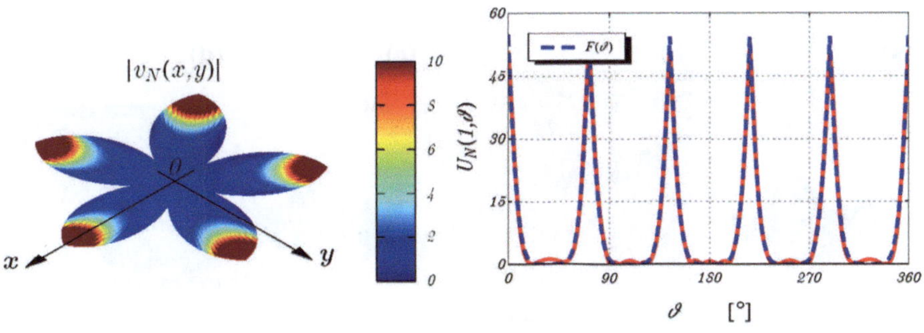

Fig. 7 Spatial distribution (*left*) and boundary values (*right*) of the partial sum $u_N(x, y)$ of order N = 20 approximating the solution of the interior Dirichlet problem for the Helmholtz equation in a corolla shaped domain

laws relate two parameters raised to some power via multiplication, while Lamé curves do this via addition [42].

Choripetaly and sympetaly in flowers can be based on these two most elementary mathematical operations from the MATHS-Box, multiplication and addition, between a fundamental function and a constraining space. This then leads to the *XYZ* system, which is a geometrical system, not only combinatorial. The use of only the two most elementary mathematical operations from the MATHS-Box leads immediately to geometric and arithmetic means (*GM* and *AM*) between *CF* and *DF*, unveiling simple relations between arithmetic *AM* and geometric *GM* means between *CF* and *DF* to study flower development and evolution. These classical means are pivotal in number theory and the theory of surfaces. $GM < AM$ is the cornerstone in number theory and in the theory of surfaces, with K and H the square of the geometric and arithmetic mean of the two principal curvatures k_1 and k_2, respectively. This is summarized in Table 3 indicating also weighted *AM-GM* inequalities are possible.

In mathematical physics (minimal surfaces) $K \leq H^2$ is key. For example, natural shapes, such as soap bubbles and films are known to strive for equality in the

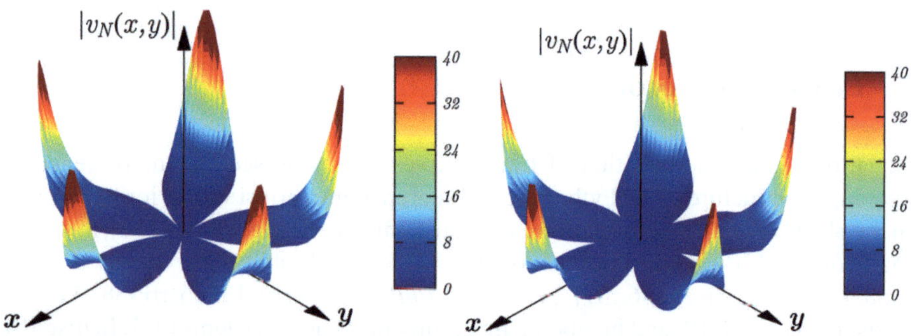

Fig. 8 Spatial distribution of the partial sum $u_N(x, y)$ of order N approximating the solution of the interior Dirichlet problem for the Helmholtz equation in the flower-shaped domain. *Left* $\alpha = 0.1$; *Right* $\alpha = 0.4$. N = 20 for both cases

Fig. 9 **a**, **c** Spatial distribution of the partial sum $u_N(x, y)$ of order N approximating the solution of the interior Dirichlet problem for the Helmholtz equation in the flower-shaped domain with fusion parameter α (a: $\alpha = 0.7$; $N = 40$; c: $\alpha = 0.9$; $N = 40$). **b** Flowers of *Stapelia arenosa*. **d** Nectar disc of *Hekistocarpa* (Rubiaceae)

Table 3 Geometric, arithmetic and weighted arithmetic means in numbers, surfaces and flowers

	Geometric mean GM	Arithmetic mean AM	Weighted arithmetic mean WAM	Relations
Numbers	$\sqrt{a.b}$	$(a+b)/2 =$	$w_1.a + w_2.b$	$GM \le AM$
Surfaces	$\sqrt{(\kappa_1\kappa_2)} = \sqrt{K}$	$(\kappa_1+\kappa_2)/2 = H$	$\kappa_1 cos^2\varphi + \kappa_2 sin^2\varphi$	$K \le H^2$
Flowers	$\sqrt{(DF.CF)}$	$(DF+CF)/2$	$\alpha CF + (1-\alpha)DF$	$DF.CF \le ((DF+CF)/2)^2$

classical inequalities between intrinsic and extrinsic curvatures like $K \le H^2$ (with K Gaussian curvature and H the mean curvature). Similar relations can now be used to study flower morphology based on multiplication and addition of CF and DF. The constraining polygons CF define a curvature, not based on the classical circle, but taking into account the specific geometry of the flower and the original symmetry of the flower bud. The use of a weighted addition, in which both functions contribute, was also used by John Nash in his groundbreaking work in cooperative games [43].

The extension of the theory of minimal surfaces to incorporate these novel transformations has already been achieved [32]. In particular, the extension of constant mean curvature surfaces describing the shape of soap bubbles and films, to the anisotropic case will allow for understanding the development of natural shapes like flowers as a process of minimizing tension based on the morphology defined by our *XYZ* system. Stress distribution and minimization are indeed observed in Asclepiad corollas and petal fusion results in an area increase and a more stable central area in the flower. This stability is absolutely necessary for the precise formation of complex novel organs through inter- and intra-whorl fusion events in large flowers.

8 Fourier Projection Method as a General Method for BVP

This stability of fused petals results immediately from our exact, analytical solutions of the Helmholtz equation. This equation is used in the study of many problems of mathematical physics involving time and space and a range of problems can be reduced to (coupled) Helmholtz equations such as plate vibrations and Turing's pattern formation.

The Helmholtz equation relates to many problems of mathematical physics with partial differential equations involving time and space, and is used to study problems in electromagnetism, acoustics and vibrations etc. Regular solutions of this equation are metaharmonic functions [44]. Many other problems can be reduced to the Helmholtz equation: for example, plate vibrations involving a fourth order Laplacian, can be solved using two independent Helmholtz equations. A hyperbolic equation, like the wave equation, can be transformed into Helmholtz using the Laplace transform. Here we used it for plate vibrations as a system of two Helmholtz equations. Similar systems are also found in patterns formation in biological specimen, so-called Turing systems. The solution of this equation, for various boundary conditions can now be expressed based on the Fourier projection method. This provides an analytical method for combining geometry with molecular and calculus-based biophysical paradigms as conjectured by Paul B. Green [45].

The original memoir of Jean-Baptiste Fourier in 1807 [46] aimed at providing a solution to a specific problem in mathematical physics, the heat equation, and he showed that this particular problem could be solved using classical trigonometric functions. Despite the widespread use of Fourier series in science and engineering it remained unproven for a very long time whether the idea of Fourier to approximate functions with any degree of precision using Fourier series was valid. It was only in 1966 that Lennart Carleson [47] could prove Luzin's conjecture, namely the point-wise convergence of Fourier series, almost everywhere, except in sets of measure zero.

This settled a long-standing problem, but it did not complete Fourier's project of mathematical physics. The original problem was a quadruple of a BVP, boundary conditions BC (Dirichlet, Neumann, Robin), a given boundary function BF, and a domain D, and the original solution in terms of Fourier series (i.e. of trigonometric functions defined on the classical circle). It was given on a circular domain, and for the next two centuries analytical solutions remained possible only for a few domains (e.g. in the equilateral triangle, by Gabriel Lamé). A variety of methods were developed, very few analytic, most numerical, but none of these methods could use the classical Fourier projection method [48]. It was only in 2007, exactly 200 years after Fourier's original memoir that it was shown that solutions of BVP's for any normal polar domain could be expressed in Fourier series [20]. The domains not only included normal polar planar domains, but also 3D domains and Riemann surfaces. Later these solutions were extended even to annular domains and shells.

In the stretched co-ordinate system the original domain is transformed into the unit circle or unit sphere and in such a system classical techniques like the separation of variables can be used for solving the transformed equation. Since the separation of variables is already encoded in the transformations, this becomes straightforward. Our findings are in good agreement with the theoretical results on Fourier series due to Carleson. In our methods the L^2-norm of the difference between the exact solution and its approximate value is always vanishing on the considered domain, and point-wise convergence seems to hold on the whole boundary, with the exception of a set of measure zero corresponding to the singular points for the function or its derivative. The choice for corollas with cusps shows that our methods can be used in cusped or quasi-cusped 2 and 3D domains or shells. It can thus be applied in the theory of prismatic shells and beams [44].

The solutions are very general and the technique can be used also when the functions $R(\vartheta)$ or $R(\vartheta, \varphi)$ are piecewise continuous functions and the boundary described by square integrable, not necessarily continuous functions. Our analytic solutions are based on the classical Fourier projection method. If the boundary values have wide oscillations, a larger number of terms in the relevant expansions is needed for desired numerical accuracies, but in general the order of expansion of Fourier series or spherical harmonics is reasonably low (<10).

In choripetalous flowers, petal responses to stress will remain restricted to each individual petal. In sympetalous flowers the response will be distributed in the corolla towards the periphery whereas in the center a stable zone is created (Figs. 6 and 7) as in corollas and nectar discs (Fig. 8). More fusion increases the size of the stable zone. They reveal that the first mode of vibrations of a floral drum indicates how flowers develop (Fig. 8), with a central stable core. Vibrations should be thought of as ways for the plate to get to rest at a specific equilibrium as efficiently as possible. Singling out the first mode of vibration does not come as a surprise: the architecture of plants allows for separation of vibration modes under wind stress [49] and the original idea of Turing's morphogenetic patterns is to separate specific spatial and temporal modes in reaction-diffusion systems.

This biogeometrical strategy has predictive value: in this study the flower was fixed in the center with free boundaries, but when fixed at the edges, which is physically realized when the perimeter of the corolla is stiffer than the central part, the first mode of vibration resembles the 3D deformation of the corolla as observed in a variety of bell-shaped and tubular flowers. The development of cups (*Asclepiads*) and floral tubes (e.g. *Petunia*) then becomes simply the most logical way for evolution to occur from a mathematical and physical point of view. Both the corolla as a stable disc and deformations of the corolla into floral tubes are then well-defined responses to local and global stresses, both in a long time frame (*evo*) or in a shorter one (*devo*).

9 Fast Forward Evolution: Combining Stochastic and Deterministic Techniques

A major question is how key innovations can be stabilized and lead to specialized groups. The key innovation of large flowers with fused petals of Rafflesiaceae evolved very fast within the Euphorbiaceae with small flowers [50]. The whole group of Asclepiads also stabilized the key innovation, and evolved into a very species-rich group. One way to understand this fast evolution is to compare developments in computer science, in particular the combination of stochastic (mutations in genetic algorithms) and deterministic (e.g. Levenberg–Marquardt) methods in image analysis.

Methods and algorithms to reconstruct Gielis curves and surfaces from data points and contours have been proposed, even in the case of the complete lack of prior information and various noises [51–53]. The most advanced techniques can also analyze self-intersecting curves which can be found for example in all projections in the plane of curves wound on helices or conics or shapes based on spherical products with self-intersecting cross sections (*Astrophytum*, Fig. 2). It is important to note that the detection of rotational symmetries of real and incomplete objects without prior information still a challenging and open problem. Furthermore, since symmetries are not differentiable parameters, they cannot be retrieved using any deterministic methods, which imposes additional pre-processing and symmetry extractions before shape recovery.

The most efficient techniques are based on combinations of stochastic and deterministic methods. Evolutionary algorithms are used for a very limited number of generations to rapidly determine broad solutions [54] that are then refined using deterministic methods [55]. In the first step, the stochastic method provides one family of solutions for the raw problem of determining the appropriate symmetry and approximations of the shape coefficients without any prior information. Through a careful definition and construction of evolutionary operators coupled with a cost function built upon the Euclidean distance, the algorithm very quickly eliminates shapes with incorrect symmetries to only retain a population of few approximate solutions. Because a major drawback of stochastic methods is their slow convergence, the fine-tuning of an optimal solution is obtained in the second step using the Levenberg–Marquardt algorithm [54, 55].

Consequently, this strategy allows for robust and efficient shape initialization with quick and accurate determination of a solution. By way of comparison: if only stochastic methods were used for the analysis, this would require thousands of generations for large populations, whereas now a few hundreds of generations for much smaller populations suffice. Oppositely, if only deterministic methods were used, some of the slightest variations of the initial guess, for instance due to missing data or incoherent pose estimation, might lead to local minima and/or numerical instabilities. Stochastic approaches need not be restricted to evolutionary algorithms, but other methods like particle swarm optimization *PSO* can be used to determine the optimal parameters for the optimal shape. In particular, in antennas, a new algorithm called *enhanced weighted quantum PSO* led to very fast and very efficient computations to determine optimal shapes, efficiently reducing the number of "generations" by at least one order of magnitude [56]. This can even be improved by adding deterministic optimization in the final stages.

We believe that this hybrid approach helps understanding natural evolutionary processes. Such strategies may result in very efficient strategies with short evolutionary times, combining efficient mutations with deterministic optimization strategies. Examples are the mobilization and neo-functionalization of pre-existing enzymes, large numbers of offspring or epigenetic imprinting. Once a key innovation is formed in evolution as a broad solution, rapid diversification will follow, with adaptations not only to physical conditions, but also towards biotic factors, not in the least available pollinators. In this respect orchids (monocots) and asclepiads (dicots) share the formation of a central column, pollinaria and complex olfactory patterns. Both groups are very rich in species, diverging and specialization. The key innovation is fusion into pollinaria in both, in the corolla only in Asclepiads. Notwithstanding the absence of fusion of tepals in orchids, similar biophysical principles may be the basis of deformations of the labellum in orchids.

10 Conclusion

The Stapeliad corolla has been specifically selected as a physical (stability and reproducibility) and biological system (complexity); we showed that it is an excellent biogeometrical model for the study of symmetry, synorganization and stability in development and evolution in flowers. Synorganization is found in reproductive structures in over 80% of all angiosperms and is very common in vegetative leaves and stems, irrespective whether synorganised parts arise by fusion or apoptosis. Our model can be used to understand fusion as the basis of the stunning stability and floral diversity in Asclepiadaceae and the rapid evolution of giant flowers of *Rafflesia*, *Rhizanthes* and *Sapria* within the Euphorbiaceae [50]; it may shed some light on the disproportionate number of species bearing gigantic flowers in sapromyiophilous species. Our model also captures the evolutionary development of oligomery and fusion, two major trends in the evolution of plants. It exhibits a very general applicability enabling to model any flower shape.

In a broader sense our model presents a geometrical and biophysical strategy to study flower development and evolution with symmetry, synorganization, stability and shape at the core, irrespective of the underlying molecular mechanisms like neo-functionalization [13] or micro-RNA's [30]. Our model shows that any study of flower development in Asclepiads, Rafflesiaceae and plants in general, should interpret expression patterns of *(A)BC* genes in a broader framework of choripetaly and sympetaly, fusion and synorganisation. *Asclepias* has already been targeted as a model species using next generation sequencing [57], and a specific strategy could be devised for a comparative study of *Petunia maewest*, Asclepiads and orchids unraveling the control factors and gene complexes, as well as the epigenetic regulation of these important trends in angiosperm (and angiophyte) evolution. The comparison between orchids and Asclepiads is interesting since aggregation of pollen into pollinaria in both groups is preceded in evolution by the formation of column (Orchids) and gynostegium (Asclepiads). This synorganisation has led to two groups very rich in species [58].

With major focus on the *MADS BOX* gene family in flower formation, progress in understanding fusion and synorganisation has been very slow [59, 60]. To study fusion or lack thereof in flowers *Petunia maewest* is still the major reference with reduced lateral outgrowth of initially separate primordial, failing to fuse during development [16, 61]. This process involves *HOMEOBOX* genes. *MAW* (*Petunia maewest*) encodes a member or the *WOX1* subfamily and the role of *MAW/WOX1* homologues in other plants seem to have similar phenotypes [61, 62]. This does shed some light on fusion, but a thorough study and understanding will need a multifaceted approach, including physiology, biophysics and geometry. It is extremely important for the whole field of botany to study a wide variety of plants, e.g. *Petunia* [63–65]. One must always keep a broad and general view on how nature explores every possible pathway, including fusion resulting in species rich subgroups once such innovations are stabilized in monocots and dicots [66], and how they affect evolution. The division of Asterids into Campanullids and Lamiids is largely based on the timing of petal fusion, early or late sympetaly respectively, leading to floral tubes. Also in the Rosids however, the formation of tubes is possible as Geranium, and in this case the separate petals anchor in the lower zones [67].

With our biogeometrical approach we hope to open a framework for understanding fusion in plants, helping to understand the ubiquity of synorganization in plants by fusion between different parts and organs, as a major driving force in the evolution of angiosperms. We emphasize that synorganization is a broad phenomenon. For example, the cardioid analogy applied to choripetalous and sympetalous flowers then allows for modeling a wide range of composite or palmate leaves with differing degrees of fusion, as can be observed within the genus *Acer*. In an evo-devo perspective, understanding flower formation and evolution as the sound of drums favors the idea of a co-active role of plants and pollinators in co-evolution. The combination of trigonometric functions with Gielis transformations, with the exact solution to boundary value problems in arbitrary domains, as applied here with the Helmholtz equation for floral development and morphology, can in fact be used in any field of science.

References

1. Z. Schwartz-Sommer, P. Huijser, W. Nacken, H. Saedler, H. Sommer, Genetic control of flower development by homeotic genes in Antirrhinum majus. Science **250**, 931–936 (1990)
2. E.S. Coen, E.M. Meyerowitz, The war of the whorls: genetic interactions controlling flower development. Nature **353**, 31–37 (1990)
3. A.S. Rijpkema, M. Vandenbussche, R. Koes, K. Heijmans, T. Gerats, Variations on a theme: changes in the floral ABCs in angiosperms. Semin. Cell Dev. Biol. **21**(1), 100–107 (2010)
4. B. Causier, Z. Schwartz-Sommer, B. Davies, Floral organ identity: 20 years of ABCs. Semin. Cell Dev. Biol. **21**, 73–79 (2010)
5. K. Heijmans, K. Ament, A.S. Rijpkema, J. Zethof, M. Wolters-Arts, T. Gerats, M. Vandenbussche, Redefining C and D in the Petunia ABC. Plant Cell **24**(6), 2305–2317 (2012)
6. P.J. Rudall, R.M. Bateman, Defining the limits of flowers: the challenge of distinguishing between the evolutionary products of simple versus compound strobili. Philos. Trans. R. Soc. B **365**, 397–409 (2010)
7. P.K. Endress, Origins of flower morphology. J. Exp. Zool. (Mol. Dev. Evol.) **291**, 105–115 (2001)
8. L.R. Decraene, E.F. Smets, The distribution and systematic relevance of the androecial character oligomery. Bot. J. Linn. Soc. **118**(3), 193–247 (1995)
9. L. Ronse De Craene, *Floral Diagrams* (Cambridge University Press, Cambridge, 2011)
10. P.K. Endress, Evolutionary diversification of the flowers in angiosperms. Am. J. Bot. **98**(3), 370–396 (2011)
11. F. Weberling, *Morphology of Flowers and Inflorescences* (Cambridge University Press, Cambridge, 1992)
12. P.K. Endress, *Diversity and Evolutionary Biology of Tropical Flowers* (Cambridge University Press, Cambridge, 1994)
13. M. Vandenbussche, G. Theissen, Y. Van de Peer, T. Gerats, Structural diversification and neo-functionalization during floral MADS-box gene evolution by C-terminal frameshift mutations. Nucl. Acids Res. **31**(15), 4401–4409 (2003)
14. T. Gerats, M. Vandenbussche, A model system for comparative research: Petunia. Trends Plant Sci. **10**(5), 251–256 (2005)

15. Z.H. Gonzalez-Carranza, U. Rompa, J.L. Peters, A. Bhatt, C.A. Wagstaff, A.D. Stead, J.A. Roberts, Hawaiian skirt: an F-box gene that regulates organ fusion and growth in Arabidopsis. Plant Physiol. **144**, 1370–1382 (2007)
16. M. Vandenbussche, A. Horstman, J. Zethof, R. Koes, A. Rijpkema, T. Gerats, Differential recruitment of WOX transcription factors for lateral development and organ fusion in Petunia and Arabidopsis. Plant Cell **21**, 2269–2283 (2009)
17. N.J. Vereecken, J.N. McNeil, Cheaters and liars: chemical mimicry at its finest. Can. J. Zool. **88**, 725–752 (2010). doi:10.1139/Z10-040
18. F. Alberts, U. Meve, *Illustrated Handbook of Succulent Plants: Asclepiadaceae* (Springer, Berlin, 2002)
19. J. Gielis, A generic geometric transformation that unifies a wide range of natural and abstract shapes. Am. J. Bot. **90**(3), 333–338 (2003)
20. P. Natalini, R. Patrizi, P.E. Ricci, The Dirichlet problem for the Laplace equation in a starlike domain of a Riemann surface. Numer. Algorithms **28**, 215–227 (2007)
21. D. Caratelli, P.E. Ricci, J. Gielis, The Robin problem for the Laplace equation in a three-dimensional starlike domain. Appl. Math. Comput. **218**(3), 713–719 (2011)
22. D. Caratelli, J. Gielis, I. Tavkelidze, P.E. Ricci, Fourier-Hankel solution of the Robin problem for the Helmholtz equation in supershaped annular domains. Bound. Value Probl. **2013**(113), 1–10 (2013)
23. D. Caratelli, P. Natalini, P.E. Ricci, Fourier solution of the wave equation for a star-like-shaped vibrating membrane. Comput. Math. Appl. **59**, 176–184 (2010)
24. D. Caratelli et al., *Fourier Solution of the Dirichlet Problem for the Laplace and Helmholtz Equations in Starlike Domains* (Tbilisi University Press, 2010). ISSN 1512-0511
25. J. Gielis, *Inventing the Circle* (Geniaal Press, Antwerpen, 2003)
26. S. Wang, J.Z. Pan, Integrating and querying parallel leaf shape descriptions, *International Semantic Web Conference* (Springer, Berlin, 2006), pp. 668–681
27. J. Gielis, T. Gerats, Form and shape of plants and plant organs are determined by a single geometrical superformula, in *14th International Symposium Morphology, Anatomy and Systematics of Plants, Jena*, ed. by Helwig, et al. (Program and Abstracts, 1999)
28. I. Tavkhelidze, C. Cassisa, J. Gielis, P.E. Ricci, About "Bulky" links, generated by generalized Möbius-Listing's bodies. Rendiconti Lincei-Matematica e Applicazioni **24**(1), 11–38 (2013)
29. J. Gielis, T. Gerats, A botanical perspective on modeling plants and plant shapes in computer graphics, in *International Conference on Computer, Communication and Control Technologies. Austin, Texas* (2004)
30. M. Cartolano, R. Castillo, N. Efremova, M. Kuckenberg, J. Zethof, T. Gerats, Z. Schwarz-Sommer, M. Vandenbussche, A conserved microRNA module exerts homeotic control over Petunia hybrida and Antirrhinum majus floral organ identity. Nature Genetics Published Online 24 June 2007
31. A.W. Leissa, Vibrations of Plates (1969). NASA SP-160
32. M. Koiso, B. Palmer, Rolling constructions for anisotropic Delaunay surfaces. Pac. J. Math. **234**(2), 345–378 (2008)
33. S. Zhou, Q. Li, S. Townsend, Y.M. Xoie, X. Huang, Fishnet metamaterial with double negative refractive index in blue region of visible spectrum, in *Proceedings of the SPIE 8806, Metamaterials: Fundamentals and Applications VI, 88062B* (11 September 2013). doi:10.1117/12.2035386
34. E. Moreno, M. Villada, P. Ruiz, F.G. Roldan, E.G. Marin, A new explicit and analytical model for square Gate-All-Around MOSFETs with rounded corners. Solid-State Electron. **111**, 180–187 (2015)
35. P. Bia, D. Caratelli, L. Mescia, J. Gielis, Analysis and synthesis of supershaped dielectric lens antennas. IET Microw. Antennas Propag. **9**(14), 1497–1504 (2015)
36. S. Lin, L. Zhang, G.V. Reddy, C. Hui, J. Gielis, Y. Ding, P. Shi, A geometrical model for testing bilateral symmetry of bamboo leaf with a simplified Gielis equation. Ecol. Evol. **6**(19), 6798–6806 (2016)

37. P.-J. Shi, Q. Xu, H.S. Sandhu, J. Gielis, Y.-L. Ding, H.-R. Li, X.-B. Dong, Comparison of dwarf bamboo (Indocalamus sp.) leaf parameters to determine relationship between spatial density of plants and total leaf area per plant. Ecol. Evol. (2015). doi:10.1002/ece3.1728
38. T.R. Faisal, E.M.K. Abad, N. Hristozov, D. Pasini, The impact of tissue morphology, cross-section and turgor pressure on the mechanical properties of the leaf petiole in plants. J. Bionic Eng. **7**, S11–S23 (2013)
39. T.R. Faisal, N. Hristozov, T.L. Western, A. Rey, D. Pasini, The twist-to-bend compliance of the Rheum rhabarbarum petiole: integrated computations and experiments. Comput. Methods Biomech. Biomed. Eng. 1–12 (2016)
40. P.J. Shi, J.G. Huang, C. Hui, H.D. Grissino-Mayer, J.C. Tardif, L.H. Zhai, B.L. Li, Capturing spiral radial growth of conifers using the superellipse to model tree-ring geometric shape. Front. Plant Sci. **6** (2015)
41. Q. Wei, C. Jiao, L. Guo, Y. Ding, J. Cao, J. Feng, G. Yang, et al., Exploring key cellular processes and candidate genes regulating the primary thickening growth of Moso underground shoots. New Phytol. doi:10.1111/nph.14284
42. S. Allesina, E. Sander, M.J. Smith, S. Tang, Superelliptical laws for complex networks (2013). arXiv:1309.7275v2
43. J. Nash, *Essays on Game Theory* (Princeton University Press, Princeton, 1996)
44. I. Vekua, *On Metaharmonic Functions*, vol. 14, Lecture Notes of TICMI (Tbilisi University Press, 2013)
45. P.B. Green, Expression of pattern in plants: combining molecular and calculus-based biophysical paradigms. Am. J. Bot. **86**(8), 1059–1076 (1999)
46. J. Fourier, *Analytical Theory of Heat*, Great Books of the Western World (Encyclopedia Brittanica, 1952)
47. L. Carleson, On convergence and growth of partial sums of Fourier series. Acta Math. **116**(1), 135–157 (1966)
48. G.P. Tolstov, *Fourier Series* (Dover Inc., New York, 1962)
49. M. Rodriguez, E. De Langre, B. Moulia, A scaling law for the effects of architecture and allometry on tree vibration modes suggests a biological tuning to modal compartmentalization. Am. J. Bot. **95**(12), 1523–1537 (2008)
50. L.A. Nikolov, P.K. Endress, M. Sugumaran, S. Sasirat, S. Vessabutr, E.M. Kramer, C.C. Davis, Developmental origins of the world largest flowers, Rafflesiaceae. Proc. Natl. Acad. Sci. U. S. A. **110**, 18578–18583 (2013)
51. Y. Fougerolle, A. Gribok, F. Truchetet, in *Proceedings of the International Conference on Image Processing ICIP'06, Atlanta, GA, USA* (2006), pp. 2193–2196
52. Y. Fougerolle, F. Truchetet, J. Gielis, A new potential function for self intersecting Gielis curves with rational symmetries, in *International Conference on Computer Graphics Theory and Applications GRAPP'09* (INSTICC Press, 2009), pp. 90–95
53. S. Voisin, M. Abidi, S. Foufou, F. Truchetet, Genetic algorithms for 3D reconstruction with supershapes, in *Proceedings of International Conference on Image Processing, ICIP 2009, Cairo, Egypt* (IEEE, 2009), pp. 529–532
54. J. Gielis, D. Caratelli, Y. Fougerolle, P.E. Ricci, T. Gerats, Universal natural shapes: from unifying shape description to simple methods for shape analysis and boundary value problems. PloS One **7**(9), e29324 (2012)
55. Y. Fougerolle, J.F. Gielis, F. Truchetet, A robust evolutionary algorithm for the recovery of rational Gielis curves. Pattern Recognit. **46**(8), 2078–2091 (2013)
56. L. Mescia, P. Bia, D. Caratelli, M.A. Chiapperino, O. Stukach, J. Gielis, Electromagnetic mathematical modeling of 3D supershaped dielectric lens antennas. Math. Probl. Eng. **2016**, 1–10 (2016)
57. S.C.K. Straub et al., BMC Genomics **12**, 211 (2011)
58. L.D. Harder, S.D. Johnson, Int. J. Plant Sci. **169**(1), 59–78 (2009)
59. C.D. Specht, D.G. Howarth, Adaptation in flower form: a comparative evodevo approach. New Phytol. **206**(1), 74–90 (2015)

60. J. Zhong, J.C. Preston, Bridging the gaps: evolution and development of perianth fusion. New Phytol. **208**(2), 330–335 (2015)
61. E. Costanzo, C. Trehin, M. Vandenbussche, The role of WOX genes in flower development. Ann. Bot. **114**(7), 1545–1553 (2014)
62. L. Niu, H. Lin, F. Zhang, T.W. Watira, G. Li, Y. Tang, M. Tadege, LOOSE FLOWER, a WUSCHEL-like homeobox gene, is required for lateral fusion of floral organs in Medicago truncatula. Plant J. **81**(3), 480–492 (2015)
63. M. Vandenbussche, P. Chambrier, S.R. Bento, P. Morel, Petunia, your next supermodel? Front. Plant Sci. **7** (2016)
64. A.R. van der Krol, R.G. Immink, Secrets of the world's most popular bedding plant unlocked. Nat. Plants **2**, 16082 (2016)
65. A. Bombarely, M. Moser, A. Amrad, M. Bao, L. Bapaume, C.S. Barry, M. Bucher, Insight into the evolution of the Solanaceae from the parental genomes of Petunia hybrida. Nat. Plants **2**, 16074 (2016)
66. P.K. Endress, Development and evolution of extreme synorganization in angiosperm flowers and diversity: a comparison of Apocynaceae and Orchidaceae. Ann. Bot. mcv119
67. P.K. Endress, Synorganisation without organ fusion in the flowers of Geranium robertianum (Geraniaceae) and its not so trivial obdiplostemony. Ann. Bot. **106**, 687–695 (2010)

A Note About Generalized Forms of the Gielis Formula

Johan Gielis, Pierpaolo Natalini and Paolo Emilio Ricci

Abstract We generalize the Gielis Superformula by extending the R. Chacón approach, but avoiding the use of Jacobi elliptic functions. The obtained results are extended to the three-dimensional case. Several new shapes are derived by using the computer algebra system Mathematica©.

Keywords Superellipse · Gielis superformula · Natural shapes

MSC Classification 51N05 · 65D17

1 Introduction

In a widely cited article, J. Gielis [1] introduced the formula - generalizing the Lamé one - containing several parameters

$$r(\theta) = \left[\left(\left|\frac{\cos\left(\frac{m_1\theta}{4}\right)}{\alpha}\right|^{n_1} + \left|\frac{\sin\left(\frac{m_2\theta}{4}\right)}{\beta}\right|^{n_2}\right)\right]^{-1/n_3}, \tag{1}$$

J. Gielis
Department of Bioengineering Sciences, University of Antwerp,
Groenenborgerlaan, 171, 2020 Antwerp, Belgium

P. Natalini
Dipartimento di Matematica e Fisica, Università Degli Studi Roma Tre,
Largo San Leonardo Murialdo, 1, 00146 Rome, Italy
e-mail: natalini@mat.uniroma3.it

P.E. Ricci (✉)
International Telematic University UniNettuno,
Corso Vittorio Emanuele II, 39, 00186 Rome, Italy
e-mail: paoloemilioricci@gmail.com

J. Gielis et al. (eds.), *Modeling in Mathematics*,
Atlantis Transactions in Geometry 2, DOI 10.2991/978-94-6239-261-8_8

and so flexible to describe a wide set complex shapes that can be found in nature, such as polygons, stars, spirals, flowers [2]. This is why this formula is referred in literature as the Gielis Superformula [3–5].

In a subsequent article R. Chacón [6], proposed the following equation

$$r_{I,II}(\theta) = \left[\left(\left|\frac{\cos\left(\Phi_{I,II}(\theta)\right)}{\alpha}\right|^{n_1} + \left|\frac{\sin\left(\Psi_{I,II}(\theta)\right)}{\beta}\right|^{n_2}\right)\right]^{-1/n_3}, \tag{2}$$

where $\Phi_{I,II}$ and $\Psi_{I,II}$ are suitable amplitudes of Jacobi elliptic functions. The motivation of this choice, according to the author, is because the solution of many nonlinear physical models are expressed in terms of such functions.

But, dealing with modelling geometrical shapes, this selection appears quite restrictive, since the non-linearity is restricted by the use of these particular functions. In our opinion, in order to generalize the Gielis formula, it is more simple to introduce the following more general equation:

$$r(\theta) = c(\theta)\left[\left(\frac{1}{\alpha}\left|\cos\left(\frac{m_1 f_1(\theta)}{4}\right)\right|^{n_1} + \frac{1}{\beta}\left|\sin\left(\frac{m_2 f_2(\theta)}{4}\right)\right|^{n_2}\right)\right]^{-1/n_3}, \tag{3}$$

where $\theta \in [-\pi, \pi]$, f_1, f_2 and $c(\theta)$ are continuous functions, m_1, m_2, α, β, are real parameters and n_1, n_2, n_3, integral numbers.

Remark 1 - Of course, division by 4 in the preceding formula is unnecessary, since same results can be simply obtained by changing the values m_1 and m_2. Actually, this division is only assumed in order to preserve the original form of the Gielis Superformula and the special case of Lamé curves for $m = 4$.

In this article, we characterize different sets of possible choices for the basic functions $f_1(\theta)$ and $f_2(\theta)$ and $c(t)$ - first starting from the more simple case $f_1(\theta) \equiv f_2(\theta) := f(\theta)$ - and we show the relevant curves for particular set of parameters, by using the computer algebra system Mathematica©.

In Sect. 4, we extend our results to the three dimensional case, by using similar sets of basic functions, and same or different parameters.

2 Notable Properties of the Gielis Formula

In order to emphasize the novelty of our approach with respect to preceding works, we introduce the precise notion of Pythagorean-compact description that is a further way of describing how complexity is reduced to simplicity. The circle has its most compact expression in polar coordinates. All circles can be represented by $r = R$; all circles are defined by one number R which determines size; the shape is fixed. With Pythagorean-compact we mean the description of size and shape in the same compactness of the Pythagorean Theorem, the most compact way of representing circles. The expression $r = R$ hides one important term, namely the Gielis transformation for $n = 2$ and $m = 4$. In this case $\rho(\theta)$ is not a radius with variable length dependent on the angle, but fixed since this term is equal to one.

$$\rho(\theta) = \frac{R}{\sqrt{|\cos\theta|^2 + |\sin\theta|^2}}; \qquad \frac{\rho(\theta)}{R} = 1 \tag{4}$$

Supershapes and Lamé curves in our polar representation clearly have the same compact structure as the Pythagorean Theorem, with two perpendicular ways of measuring spacing (cosine and sine), and size parameters R or A and B in terms of ellipses. The main difference is the way of spacing the coordinates on the perpendicular axes, governed by the exponents, and the angles between the axes, by the parameter m.

$$\rho(\theta) = \frac{R}{\sqrt[n_3]{\left|\frac{1}{\alpha}\cos\left(\frac{m\theta}{4}\right)\right|^{n_1} + \left|\frac{1}{\beta}\sin\left(\frac{m\theta}{4}\right)\right|^{n_2}}} \tag{5}$$

Hence, the Gielis Formula is Pythagorean-compact, and our transformations combine the compactness of the Pythagorean Theorem with its greatest advantage, the orthogonality (independence) of its variables. Within the same structure and same compactness we can define stretched radii and different spacing of angles (3) when $\frac{m\theta}{4}$ is generalized to $f(\theta)$.

Remark 2 - Note that, if we wish to describe the complexity of an object, it is not only the degree, but also the number of monomials in the polynomials, which describe the curve or object. In the case of Lamé curves, and the subset (with absolute values) of superellipses defined by $\left|\frac{x}{A}\right|^n + \left|\frac{y}{B}\right|^n = 1$, irrespective of the degree, the number of monomials is two, one in each variable. This structure is maintained for superellipsoids $\left|\frac{x}{A}\right|^n + \left|\frac{y}{B}\right|^n + \left|\frac{z}{C}\right|^n = 1$. Gielis curves and surfaces generalize Lamé's ideas for any symmetry (integer and non-integer) and for any degree (including non-integer powers and roots).

Such finite and short expansions can be named oligomials, a complete description in a few (oligo) terms. The name oligomials first appears in a text by V.I. Arnold and O. Oleinik [7], where they expressed the need for a better way to describe shape. "What is essential for the topological complexity of an object is not the degree of the equation, but rather the number of monomials that appear in the polynomial with nonzero coefficients. Thus, we have the problems of oligomials, i.e. the topology of objects specified by polynomials of arbitrarily large degree but with a restricted number of monomials".

Oligomials can be specified by polynomials of arbitrarily large degree but with a restricted number of monomials. This is the case in the Pythagorean Theorem and in Lamé and Gielis curves. Moreover, in all three cases there is an explicit relation between the monomials: they are monomials in single variables, not mixed (or using geometric means). Hence it seems that, apart from being Pythagorean-compact, in this sense our formulas are also topologically simple (and Pythagorean-simple).

Remark 3 - The Gielis formula as generalization of Lamé curves and superellipses uses transcendental functions, the trigonometric ones. Recently however, it was observed that these equations can be rewritten in terms of Chebyshev polynomials, for $x = \cos\frac{\theta}{4}$, based on Grandi curves [8].

$$\rho(x) = \frac{R}{\sqrt[n_1]{\left|\frac{1}{\alpha}T_m(x)\right|^{n_2} + \left|\frac{1}{\beta}\sqrt{1-x^2}\,U_{m-1}(x)\right|^{n_3}}} \tag{6}$$

For a suitable choice of parameters, the shapes defined by (1) can then be treated as rational functions. Moreover, instead of R in (5), we may use Chebyshev polynomials, and in this way also this family of polynomials is generalized. By using substitutions of the type $x = \cos(f(\theta))$, for reasonable choices of $f(\theta)$ such as used in Sect. 3, (6) can be generalized for (2) and (3) as well.

3 Plane Non Symmetric Curves

We start assuming $c(\theta) = c$ (constant), and sometimes $c = 1, f_1(\theta) \equiv f_2(\theta) := f(\theta)$.

In order to obtain plane non symmetric curves the following different choices for the basic function $f(\theta)$ can be performed:

3.1 Basic Increasing Function $f(\theta)$

A possible choice for $f(\theta)$ is an increasing function such that $f(-\pi) = -\pi$ and $f(\pi) = \pi$ (Fig. 1).

For example, we can choose (Fig. 2):

$$f(\theta) = \pi(\theta/\pi)^{2m+1}\,, \qquad (m = 1, 2, 3, \dots)\,,$$

$$f(\theta) = 2\arcsin(\theta/\pi)\,,$$

or,

$$f(\theta) = \pi\sin(\theta/2)\,,$$

$$f(\theta) = \pi\tan(\theta/4)\,.$$

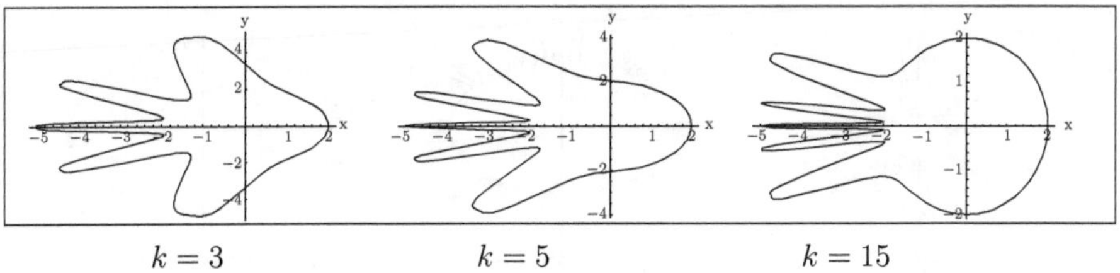

Fig. 1 $c(\theta)=1, \alpha=2, \beta=5, m_1=m_2=10, n_1=3, n_2=3, n_3=3, f_1(\theta)=f_2(\theta)=\pi(\theta/\pi)^k$, $\theta \in [-\pi, \pi]$

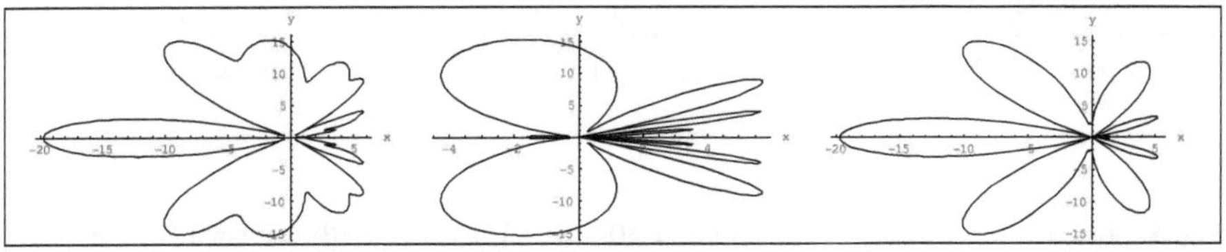

Fig. 2 $c(\theta)=2(\theta^2-\pi^2), m_1=m_2=19, n_1=14, n_2=11, n_3=-9, f_1(\theta)=2\arcsin(\theta/\pi)$, $f_2(\theta)=\pi(\theta/\pi)^3, \theta \in [-\pi, \pi]$

3.2 Closed Curves

Note that in order to ensure the curve is closed, it is sufficient to impose the condition: $f(-\pi)=f(\pi)$. In this case, the curve will intersect the real negative axis, assuming the same value on it. In some case, the curve will pass through the origin (Fig. 3).

For example, we can choose:

$$f(\theta)=\sin^{2m+1}(n\theta), \quad (n \in \mathbf{N}, m=0,1,2,\dots),$$

or

$$f(\theta)=\cos^{2m+1}(n\theta), \quad (n \in \mathbf{N}, m=0,1,2,\dots).$$

Furthermore, it is possible to assume a trigonometric polynomial vanishing in $-\pi$ and in π, such as (Fig. 4):

$$f(\theta)=A\sin\theta+B\cos 2\theta,$$

but of course there are infinite many types of these (Fig. 5).

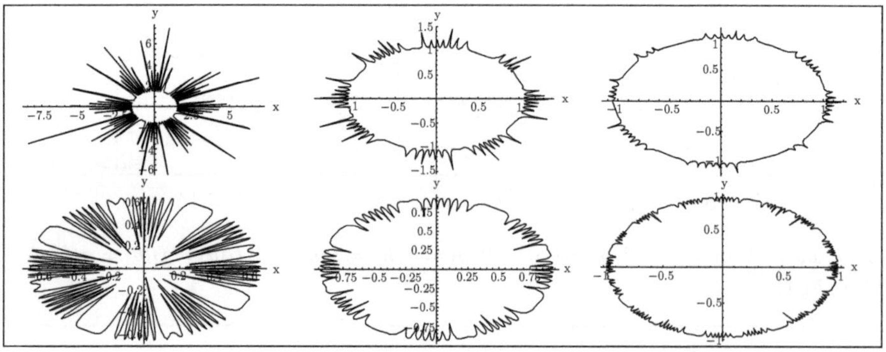

$n_3 = 5$ $n_3 = 31$ $n_3 = 75$

$n_3 = -5$ $n_3 = -31$ $n_3 = -75$

Fig. 3 $c(\theta) = 1$, $\alpha = 2$, $\beta = 7$, $m_1 = m_2 = 30$, $n_1 = 3$, $n_2 = 3$, $f_1(\theta) = 3\sin^3(2|\theta - \pi/3|)$, $f_2(\theta) = 2\cos^2[3(\theta + \pi/2)]$, $\theta \in [-\pi, \pi]$

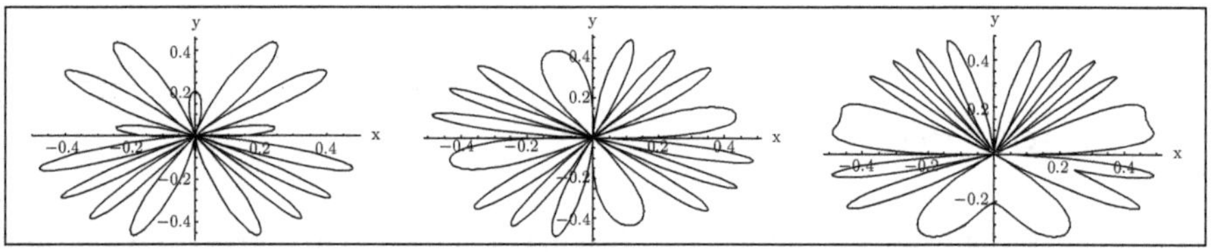

$k = 1$ $k = 2$ $k = 3$

Fig. 4 $c(\theta) = 1$, $\alpha = 2$, $\beta = 173$, $m_1 = m_2 = 4$, $n_1 = 2$, $n_2 = 2$, $n_3 = -2$, $f_1(\theta) = 3\sin(k\theta) + 5\cos(2\theta)$, $f_2(\theta) = 3\cos(\theta) + 5\sin(2\theta)$, $\theta \in [-\pi, \pi]$

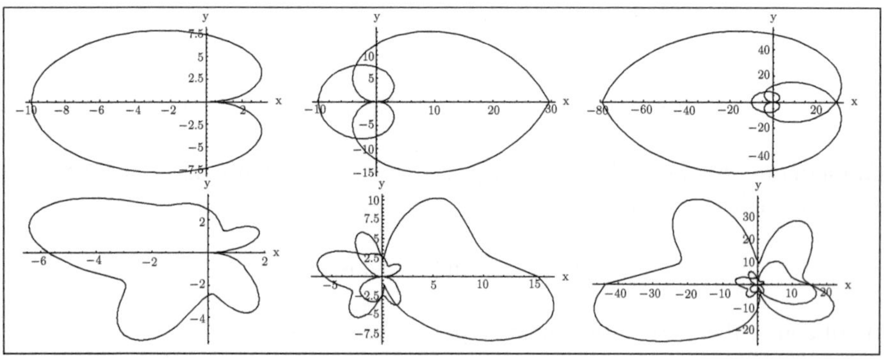

$\theta \in [-\pi, \pi]$ $\theta \in [-2\pi, 2\pi]$ $\theta \in [-3\pi, 3\pi]$

Fig. 5 - (First line) $c(\theta) = 1$, $\alpha = \beta = 1$, $m_1 = m_2 = 4$, $n_1 = 2$, $n_2 = 2$, $n_3 = -3$, $f_1(\theta) = f_2(\theta) = (\theta^2 - \pi^2)(\theta/\pi)^3$, - (Second line) $c(\theta) = 2(\theta^2 - \pi^2)$, $\alpha = 12$, $\beta = 3$, $m_1 = m_2 = 3$, $n_1 = 2$, $n_2 = 3$, $n_3 = -3$, $f_1(\theta) = 3\,|\theta|$, $f_2(\theta) = \sin(2\theta) + 3\cos(\theta)$

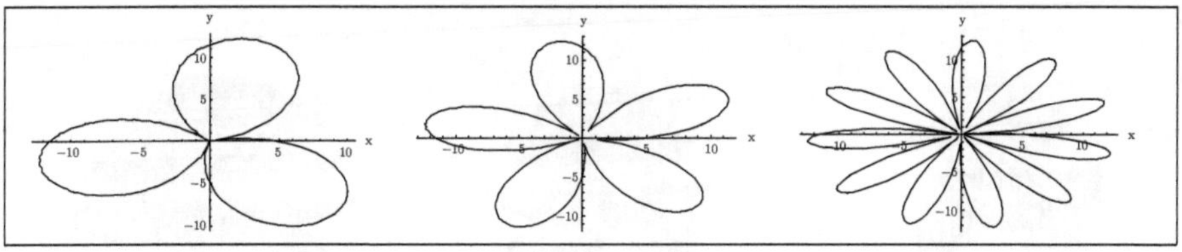

$m_2 = 3$ $m_2 = 5$ $m_2 = 11$

Fig. 6 $c(\theta) = 25, \alpha = 2, \beta = 7, m_1 = 4, n_1 = 2, n_2 = 3, n_3 = -2, f_1(\theta) = |2\theta + \pi/2|, f_2(\theta) = |3\theta + \pi/2|,\ \theta \in [-\pi, \pi]$

3.3 Curves Passing Through the Origin

Starting from a given $r(\theta)$ with $c(\theta) = c$ (constant), and changing c into $c(\theta) = (|\theta| - \pi)$ or $c(\theta) = (\theta^2 - \pi^2)$, we obtain a new function $\tilde{r}(\theta) := (|\theta| - \pi)r(\theta)$, or $\tilde{r}(\theta) := (\theta^2 - \pi^2)r(\theta)$, such that the considered function will be obliged to pass through the origin (Fig. 6).

We show a particular example of this procedure, assuming different functions for $f_1(\theta)$ and $f_2(\theta)$.

4 The Three Dimensional Case

An extension to the three dimensional case can be achieved by considering, in general, two different forms (i.e. with different basic functions and parameters) of equation (3), $r_1(\theta)$ and $r_2(\theta)$, by means of which the following parametric representation can be obtained:

$$\begin{cases} x = r_1(\theta)\cos\theta \cdot r_2(\phi)\cos\phi \\ y = r_1(\theta)\sin\theta \cdot r_2(\phi)\cos\phi \\ z = r_2(\phi)\sin\phi\,, \end{cases} \tag{7}$$

where $\phi \in [-\pi/2, \pi/2]$ is the *latitude* and $\theta \in [-\pi, \pi]$ is the *longitude*.

We show here a few examples of surfaces that can be found in such a way (Figs. 7, 8, 9, 10, 11 and 12).

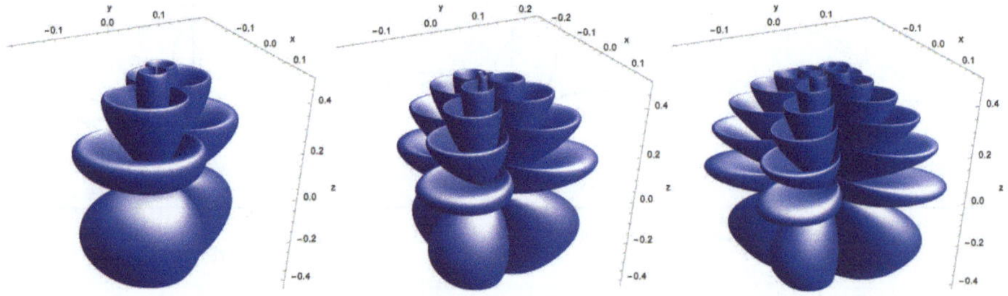

$m_1 = m_2 = 2$ $m_1 = m_2 = 3$ $m_1 = m_2 = 4$

Fig. 7 $c(\theta) = 1,\ \alpha = 3,\ \beta = 4,\ n_1 = 2,\ n_2 = 8,\ n_3 = -3, f_1(\theta) = |2\theta + \pi/2|,\ f_2(\theta) = |5\theta + \pi/2|,\ f_3(\phi) = |2\phi + \pi/2|^2,\ f_4(\phi) = |5\phi + \pi/2|^2,\ \ \theta \in [0, 2\pi],\ \phi \in [-\pi/2, \pi/2]$

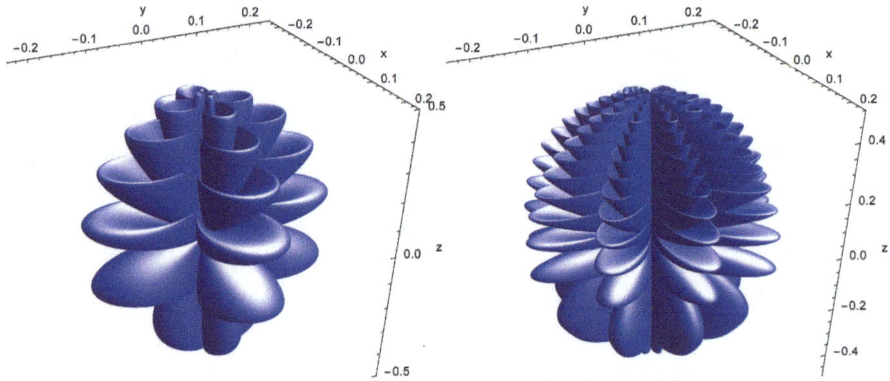

$h_i = 1,\ (i = 1, \ldots, 4)$ $h_1 = h_3 = 2; h_2 = h_4 = 3$

Fig. 8 $c(\theta) = 1,\ \alpha = 3,\ \beta = 4,\ m_1 = m_2 = 8,\ n_1 = 2,\ n_2 = 8,\ n_3 = -3, f_i(\theta) = |h_i\theta + \pi/2|,\ i = 1, 2,\ f_j(\phi) = |h_j\phi + \pi/2|^2,\ j = 3, 4,\ \ \theta \in [0, 2\pi],\ \phi \in [-\pi/2, \pi/2]$

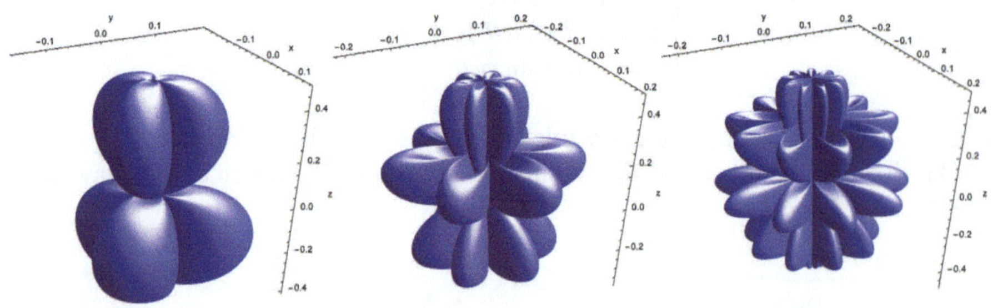

$m_1 = m_2 = 3$ $m_1 = m_2 = 6$ $m_1 = m_2 = 9$

Fig. 9 $c(\theta) = 1,\ \alpha = 3,\ \beta = 4,\ n_1 = 2,\ n_2 = 8,\ n_3 = -3,\ \ f_1(\theta) = |2\theta + \pi/2|,\ f_2(\theta) = |3\theta + \pi/2|^2,\ f_3(\phi) = |2\phi + \pi/2|,\ f_4(\phi) = |3\phi + \pi/2|^2,\ \ \theta \in [0, 2\pi],\ \phi \in [-\pi/2, \pi/2]$

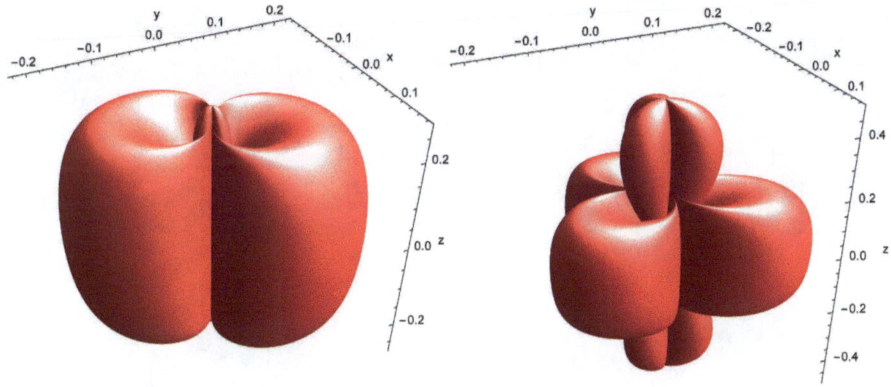

$m_1 = m_2 = 4$ $m_1 = m_2 = 6$ $m_1 = m_2 = 8$

Fig. 10 $c(\theta) = 1, \alpha = 3, \beta = 4, n_1 = 2, n_2 = 8, n_3 = -3, f_1(\theta) = |\theta - \pi/3|, f_2(\theta) = |3\theta|, f_3(\phi) = 2\cos^2(\phi + \pi/2), f_4(\phi) = |3\phi|, \theta \in [0, 2\pi], \phi \in [-\pi/2, \pi/2]$

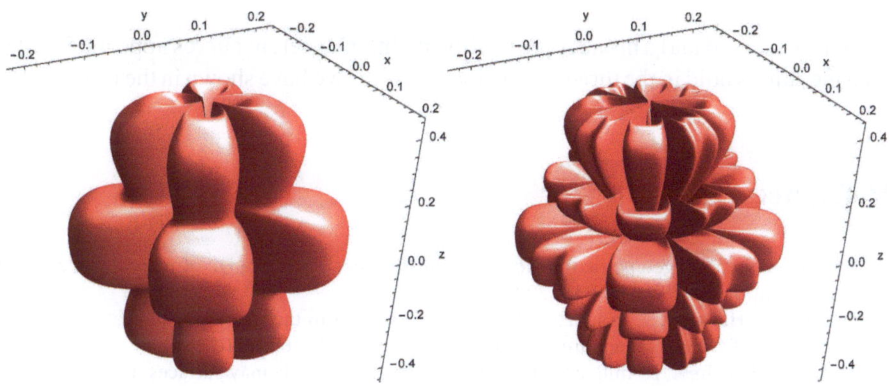

$m_1 = m_2 = 2$ $m_1 = m_2 = 5$

Fig. 11 $c(\theta) = 1, \alpha = 3, \beta = 4, n_1 = 2, n_2 = 8, n_3 = -3, f_1(\theta) = 3\sin^3|2(\theta - \pi/3)|, f_2(\theta) = |3\theta|, f_3(\phi) = 2\cos^2|3(\phi + \pi/2)|, f_4(\phi) = |3\phi|, \theta \in [0, 2\pi], \phi \in [-\pi/2, \pi/2]$

5 Conclusion

We have shown that the generalization of the Gielis superformula considered by R. Chacón [6] is not unique, but on the contrary, it is possible to obtain extensions of the superformula in an infinite way, by changing θ into suitable functions of θ, such as $f_1(\theta)$ or $f_1(\theta)$, in the original Gielis equation, provided that these functions satisfy suitable properties as reported in the preceding sections. The nonlinearity appearing

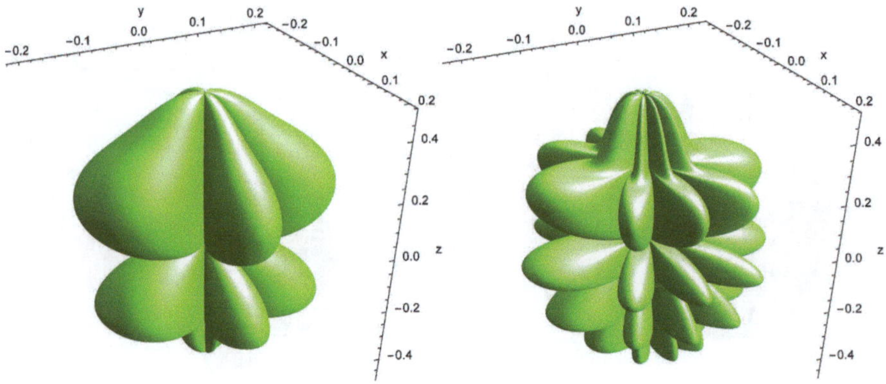

$m_1 = m_2 = 4$ $m_1 = m_2 = 8$

Fig. 12 $c(\theta) = 1, \alpha = 3, \beta = 4, n_1 = 2, n_2 = 8, n_3 = -3, f_1(\theta) = 3\sin\theta, f_2(\theta) = |2\theta|, f_3(\phi) = 3\sin\phi + 2\cos\phi, f_4(\phi) = 2|\phi|^2,\ \theta \in [0, 2\pi],\ \phi \in [-\pi/2, \pi/2]$

in [6] is preserved and a broader possibility to obtain different curves appears. Similar considerations hold in the three dimensional case, as we have shown in the last section.

References

1. J. Gielis, A generic geometric transformation that unifies a wide range of natural and abstract shapes. Am. J. Bot. **90**, 333–338 (2003)
2. J. Gielis, S. Haesen, L. Verstraelen, Universal shapes: from the supereggs of Piet Hein to the cosmic egg of Georges Lemaître. Kragujevac J. Math. **28**, 55–67 (2005)
3. M. Koiso, B. Palmer, Rolling constructions for anisotropic Delaunay surfaces. Pacific J. Math. **2**, 345–378 (2008)
4. D. Caratelli, P. Natalini, P.E. Ricci, Fourier solution of the wave equation for a star-like-shaped vibrating membrane. Comput. Math. Appl. **59**, 176–184 (2010)
5. M. Matsuura, Gielis' superformula and regular polygons. J. Geom. **106**(2), 383–403 (2015)
6. R. Chacón, Modeling natural shapes with a simple nonlinear algorithm. Int. J. Bifurc. Chaos **16**, 2365–2368 (2006)
7. V.I. Arnold, O.A. Oleinik, Topology of real algebraic manifolds. Moscow Univ. Bull. **34**, 5–17 (1979)
8. J. Gielis, D. Caratelli, P.E. Ricci, T. Gerats, The common descent of biological shape description and special functions (submitted)

Avoiding Higher Matrix Powers in the Solution of Linear Dynamical Systems

Pierpaolo Natalini and Paolo Emilio Ricci

Abstract We show that, using Lucas polynomials of the second kind, it is possible to write down explicitly the solution of linear dynamical systems - both in the discrete and continuous case - avoiding higher matrix powers. This improves the computational complexity of the algorithms usually described in literature.

Keywords Matrix powers · Linear dynamical systems · Exponential matrix · Lucas polynomials of the second kind

AMS Classification: 34A30 · 33C45 · 15A15

1 Introduction

Even in recent books (see e.g. [1, 2]), the solution of linear dynamical systems, both in the discrete or continuous time case, is expressed by using all powers of the considered matrix $\mathcal{A}_{r\times r}$. In this article, we propose an alternative method, based on recursion, using the $F_{k,n}$ functions, which are essentially linked to Lucas polynomials of the second kind [3] (i.e. the basic solution of a homogeneous linear recurrence relation with constant coefficients [4, 5]), and to the multi-variable Chebyshev polynomials [6].

We start recalling these functions and their connections with powers of a matrix [7].

The use of matrix function representation (see e.g. [8]) for the exponential function, and of the Riesz–Fantappiè formula gives us the possibility to use powers of the considered matrix up (at most) the order $r - 1$. This is a trivial consequence

P. Natalini (✉)
Department of Mathematics and Physics, Roma Tre University,
Largo San Leonardo Murialdo, 1, 00146 Rome, Italy
e-mail: natalini@mat.uniroma3.it

P.E. Ricci
International Telematic University UniNettuno,
Corso Vittorio Emanuele II, 39, 00186 Rome, Italy
e-mail: paoloemilioricci@gmail.com

J. Gielis et al. (eds.), *Modeling in Mathematics*,
Atlantis Transactions in Geometry 2, DOI 10.2991/978-94-6239-261-8_9

of the Cayley–Hamilton theorem, and should be used, in our opinion, to reduce the computational cost of solutions.

1.1 Recalling $F_{k,n}$ Functions and Lucas Polynomials of the Second Kind

Consider the $(r+1)$-terms homogeneous linear bilateral recurrence relation with (real or complex) constant (with respect to n) coefficients u_k $(k = 1, 2, \ldots, r)$ where $u_r \neq 0$:

$$X_n = u_1 X_{n-1} - u_2 X_{n-2} + \cdots + (-1)^{r-1} u_r X_{n-r}\,, \qquad (n \in \mathbf{Z}). \tag{1}$$

Supposing the coefficients vary, its solution is given by every bilateral sequence $\{X_n(u_1, u_2, \ldots, u_r)\}_{n\in\mathbf{Z}}$ such that $r+1$ consecutive terms satisfy Eq. (1).

A basis for the r-dimensional vectorial space $\mathcal{V}_r$ of solutions is given by the functions $F_{k,n} = F_{k,n}(u_1, u_2, \ldots, u_r)$, $(k = 1, 2, \ldots, r\,,\ n \geq -1)$, defined by the initial conditions below:

$$\begin{array}{llll} F_{1,-1} = 0 & F_{1,0} = 0 & \ldots & F_{1,r-2} = 1, \\ F_{2,-1} = 0 & F_{2,0} = 1 & \ldots & F_{2,r-2} = 0, \\ \ldots\ \ldots & \ldots\ \ldots & \ldots & \ldots\ \ldots \\ F_{r,-1} = 1 & F_{r,0} = 0 & \ldots & F_{r,r-2} = 0. \end{array} \tag{2}$$

Since $u_r \neq 0$, the $F_{k,n}$ functions can be defined even if $n < -1$, by means of the positions:

$$F_{k,n}(u_1, \ldots, u_r) = F_{r-k+1,-n+r-3}\left(\frac{u_{r-1}}{u_r}, \ldots, \frac{u_1}{u_r}, \frac{1}{u_r}\right), \qquad (k = 1, \ldots, r;\ n \in \mathbf{Z})\,. \tag{3}$$

Therefore, assuming the initial conditions $X_j\,,\ X_{j+1}\,,\ \ldots\,,\ X_{j+r-1}$, the *general solution* of the recurrence (1), $\forall n \in \mathbf{Z}$, is given by

$$X_{n+j+1} = X_j F_{r,n} + X_{j+1} F_{r-1,n} + \cdots + X_{j+r-1} F_{1,n}\,. \qquad (n \in \mathbf{Z})\,. \tag{4}$$

Remark – It is worth to recall that a basis for the $\mathcal{V}_r$ space is usually obtained starting from the roots of the characteristic equation

$$z^r - u_1 z^{r-1} + \cdots + (-1)^r u_r = 0\,, \qquad (r \in \mathbf{Z})\,. \tag{5}$$

If the roots $z_1, \ldots, z_r$, of (5) are all distinct, then the general solution of (1) is given by

$$X_n = C_1 z_1^n + C_2 z_2^n + \cdots + C_r z_r^n\,, \qquad (n \in \mathbf{Z})\,. \tag{6}$$

If the multiplicity of a root, for example of z_1, is $m_1 > 1$, i.e. $z_1 = z_2 = \cdots = z_{m_1}$, then, the linear combination appearing inside (6)

$$C_1 z_1^n + C_2 z_2^n + \cdots + C_{m_1} z_{m_1}^n$$

must be substituted by

$$\left(C_1 + n\,C_2 + \cdots + n^{m_1-1}\,C_{m_1}\right) z_1^n$$

and similar changes must be done for other multiple roots.

Note that this implies the knowledge of roots, whereas the use of $F_{k,n}$ is independent of that. Furthermore, the solution (4) does not depend of the multiplicity of roots.

An important result, originally stated by É Lucas [3] in the case $r = 2$, is given by the equations

$$\begin{cases} F_{1,n} = u_1 F_{1,n-1} + F_{2,n-1} \\ F_{2,n} = -u_2 F_{1,n-1} + F_{3,n-1} \\ \cdots\cdots\cdots\cdots\cdots\cdots\cdots\cdots \\ F_{r-1,n} = (-1)^{r-2} u_{r-1} F_{1,n-1} + F_{r,n-1} \\ F_{r,n} = (-1)^{r-1} u_r F_{1,n-1} \end{cases} \tag{7}$$

showing that all $\{F_{k,n}\}_{n\in\mathbf{Z}}$ functions are expressed through the only bilateral sequence $\{F_{1,n}\}_{n\in\mathbf{Z}}$.

Therefore, we assume the following

Definition – The bilateral sequence $\{F_{1,n}\}_{n\in\mathbf{Z}}$, solution of (1) corresponding to the initial conditions:

$$F_{1,-1} = 0\,, \quad F_{1,0} = 0\,, \quad \ldots\,, \quad F_{1,r-3} = 0\,, \quad F_{1,r-2} = 1\,, \tag{8}$$

is called the *fundamental solution* of (1) (“fonction fondamentale” by É. Lucas [3]).

Putting

$$F_{1,n}(u_1, \ldots, u_r) =: \Phi_n(u_1, \ldots, u_r) = \Phi_n\,, \qquad (n \in \mathbf{Z})\,. \tag{9}$$

For $n \geq -1$, the $\Phi_n(u_1, \ldots, u_r)$ functions are called in literature [4] *generalized Lucas polynomials of the second kind* (in r variables).
Remark – Note that, for $r = 2$, $u_2 = 1$, putting $u_1 = x$, we find

$$\Phi_n(u_1, 1) = \Phi_n(x, 1) \equiv U_n\left(\frac{x}{2}\right), \qquad (n \in \mathbf{N}_0),$$

where $\{U_n(x)\}_{n \in \mathbf{N}_0}$ are the second kind Chebyshev polynomials. Therefore, for $r \geq 3$, putting $u_r = 1$, an $(r-1)$-variable extension of the second kind Chebyshev polynomials is found:

$$\Phi_n(u_1, \ldots, u_{r-1}, 1) =: U_n^{(r-1)}(u_1, \ldots, u_{r-1}), \qquad (n \in \mathbf{N}_0).$$

Several properties of these polynomials can be found in literature (see e.g. R. Lidl - C. Wells [9], R. Lidl [10], T. Koornwinder [10–12], M. Bruschi - P.E. Ricci [6], K.B. Dunn - R. Lidl [13], R.J. Beerends [14]).

2 Matrix Powers Representation

In preceding articles [5, 7], the following result is proved:
Theorem – Given an $r \times r$ matrix $\mathcal{A}$, putting by definition $u_0 := 1$, and denoting by

$$P(\lambda) := \det(\lambda\mathcal{I} - \mathcal{A}) = \sum_{j=0}^{r} (-1)^j u_j \lambda^{r-j} \tag{10}$$

its characteristic polynomial (or possibly its minimal polynomial, if this is known), the matrix powers $\mathcal{A}^n$, with integral exponent n, are given by the equation:

$$\begin{aligned} \mathcal{A}^n = F_{1,n-1}(u_1, \ldots, u_r)\mathcal{A}^{r-1} + F_{2,n-1}(u_1, \ldots, u_r)\mathcal{A}^{r-2} + \\ + \cdots + F_{r,n-1}(u_1, \ldots, u_r)\mathcal{I}, \end{aligned} \tag{11}$$

where the functions $F_{k,n}(u_1, \ldots, u_r)$ are defined in Sect. 1.

Moreover, if $\mathcal{A}$ is not singular, i.e. $u_r \neq 0$, Eq. (11) still works for negative integers n, assuming the definition (3) for the $F_{k,n}$ functions.

It is worth to recall that the knowledge of eigenvalues is equivalent to that of invariants, since the second ones are the elementary symmetric functions of the first ones.

Remark – Note that, as a consequence of the above result, the higher powers of matrix $\mathcal{A}$ are always expressible in terms of the lower ones (at most up to the dimension of $\mathcal{A}$).

3 Matrix Functions Representation

It is well known that an analytic function f of a matrix $\mathcal{A} = \mathcal{A}_{r \times r}$, i.e. $f(\mathcal{A})$, is the matrix polynomial obtained from the scalar polynomial interpolating the function f on the eigenvalues of $\mathcal{A}$ (see e.g. the Gantmacher book [8]), however, in many books (see e.g. [1]), the series expansion

$$\exp \mathcal{A} = \sum_{n=0}^{\infty} \frac{\mathcal{A}^n}{n!} \tag{12}$$

is assumed for defining (and computing) $\exp(\mathcal{A})$. So, apparently, the series expansion for the exponential of a matrix is "hard to die".

Let $S_{\mathcal{A}} = \{\lambda_1, \lambda_2, \ldots, \lambda_r\}$ the spectrum of $\mathcal{A}$. Denoting by

$$\Pi_{r-1}(x) = \alpha_0 + \alpha_1 x + \cdots + \alpha_{r-1} x^{r-1}$$

the polynomial interpolating $\exp(x)$ on $S_{\mathcal{A}}$, i.e. such that: $\Pi_{r-1}(\lambda_i) = \exp(\lambda_i)$, $(i = 1, 2, \ldots, r)$, then

$$\exp \mathcal{A} = \alpha_0 \mathcal{I} + \alpha_1 \mathcal{A} + \cdots + \alpha_{r-1} \mathcal{A}^{r-1}. \tag{13}$$

If the eigenvalues are all distinct, Π_{r-1} coincides with the Lagrange interpolation polynomial and (13) is the Lagrange–Sylvester formula. In case of multiple eigenvalues, Π_{r-1} is the Hermite interpolation polynomial, and (13) reduces to Arthur Buchheim's formula, generalizing the preceding one.

This avoids the use of higher powers of $\mathcal{A}$ in the Taylor expansion (12). In any case, the possibility to write $\mathcal{A}^h$, $(h \geq 2)$, in an easy block form, requires not only the knowledge of the spectrum, but even the Jordan canonical form of $\mathcal{A}$. It is necessary to compute the eigenvectors and moreover the principal vectors, if $\mathcal{A}$ is defective. A known machinery which implies a lot of computations.

3.1 The Riesz–Fantappiè Formula

Denoting by $v_1(\lambda), \ldots v_r(\lambda)$ the invariants of the matrix $(\lambda \mathcal{I} - \mathcal{A})^{-1}$, their connections with the invariants of $\mathcal{A}$ are given by the equations:

$$
\begin{aligned}
v_1 &:= v_1(\lambda) = \binom{r}{1}\lambda - \binom{r-1}{0}u_1\,,\\
v_2 &:= v_2(\lambda) = \binom{r}{2}\lambda^2 - \binom{r-1}{1}u_1\lambda + \binom{r-2}{0}u_2\,,\\
&\cdots\cdots\cdots\cdots\cdots\cdots\cdots\cdots\cdots\cdots\cdots\cdots\cdots\cdots \qquad (14)\\
v_k &:= v_k(\lambda) = \sum_{j=0}^{k}(-1)^j\binom{r-j}{k-j}u_j\lambda^{k-j}\,,\\
&\cdots\cdots\cdots\cdots\cdots\cdots\cdots\cdots\cdots\cdots\cdots\cdots\cdots\cdots\\
v_r &:= v_r(\lambda) = P(\lambda)\,.
\end{aligned}
$$

Therefore, as it is shown in [5], we have the theorem:
Theorem – Under the hypotheses and definitions considered above, the resolvent matrix $(\lambda\mathcal{I}-\mathcal{A})^{-1}$ can be represented as

$$
(\lambda\mathcal{I}-\mathcal{A})^{-1} = \frac{1}{P(\lambda)}\sum_{h=0}^{r-1}\left[\sum_{j=0}^{r-h-1}(-1)^j u_j\lambda^{r-h-j-1}\right]\mathcal{A}^h\,. \qquad (15)
$$

Then, by the Riesz–Fantappiè formula, we recover the classical result:
Theorem – If $f(\lambda)$ is a holomorphic function in the domain $\mathcal{C}$ and denoting by $D \subset \mathcal{C}$ (with $\overline{D} \subset \mathcal{C}$) a closed set whose boundary is a piecewise simple Jordan contour γ encompassing the spectrum $S_\mathcal{A}$ of $\mathcal{A}$, the matrix function $f(\mathcal{A})$ can be represented by:

$$
\begin{aligned}
f(\mathcal{A}) &= \frac{1}{2\pi i}\oint_\gamma f(\lambda)(\lambda\mathcal{I}-\mathcal{A})^{-1}d\lambda = \\
&= \sum_{h=0}^{r-1}\left[\frac{1}{2\pi i}\sum_{j=0}^{r-h-1}(-1)^j u_j\oint_\gamma \frac{f(\lambda)\lambda^{r-h-j-1}}{P(\lambda)}\,d\lambda\right]\mathcal{A}^h\,.
\end{aligned} \qquad (16)
$$

In particular:

$$
e^{\mathcal{A}} = \sum_{h=0}^{r-1}\left[\frac{1}{2\pi i}\sum_{j=0}^{r-h-1}(-1)^j u_j\oint_\gamma \frac{e^\lambda\lambda^{r-h-j-1}}{P(\lambda)}\,d\lambda\right]\mathcal{A}^h\,. \qquad (17)
$$

Remark – If the eigenvalues of $\mathcal{A}$, are known, Eq. (16), by the residue theorem, gives back the Lagrange–Sylvester representation. However, for computing the integrals appearing in Eq. (16) it is sufficient the knowledge of a circle D, ($\gamma := \partial D$), containing the spectrum of $\mathcal{A}$ (by using the Gerschgorin theorem, and then knowing only the entries of $\mathcal{A}$, without computing its eigenvalues). Therefore, this approach is computationally more convenient with respect to the Lagrange–Sylvester formula.

4 The Discrete Time Case

Theorem – Consider the dynamical problem for the homogeneous linear recurrence system

$$\begin{cases} Z_{i+1} = \mathcal{A} \cdot Z_i \\ Z_{i=0} = Z_0 \,, \end{cases} \tag{18}$$

where

$$\mathcal{A} = \mathcal{A}_{r \times r} = \begin{pmatrix} a_{1,1} \,, & \dots \,, & a_{1,r} \\ \dots \,, & \dots \,, & \dots \\ a_{r,1} \,, & \dots \,, & a_{r,r} \end{pmatrix} .$$

Let

$$Z_n = ({z_1}^{(n)}, {z_2}^{(n)}, \dots, {z_r}^{(n)})^T \,,$$

$$Z_0 = ({z_1}^{(0)}, {z_2}^{(0)}, \dots, {z_r}^{(0)})^T \,,$$

$$Z_1 = ({z_1}^{(1)}, {z_2}^{(1)}, \dots, {z_r}^{(1)})^T = \mathcal{A} \cdot Z_0 \,,$$

$$\dots \quad \dots \quad \dots$$

$$Z_{r-1} = ({z_1}^{(r-1)}, {z_2}^{(r-1)}, \dots, {z_r}^{(r-1)})^T = \mathcal{A} \cdot Z_0^{r-2} \,,$$

denote by $u_1 := tr(\mathcal{A})$, $u_2, \dots, u_r := det(\mathcal{A})$ the invariants of $\mathcal{A}$, and recall the generalized Lucas polynomials $F_{k,n}(u_1, \dots, u_r)$, $(k = 1, 2, \dots, r \,,\; n \geq -1)$, defined in Sect. 1.

Define the vector

$$\mathcal{F}_{n-1}(u_1, \dots, u_r) := \left(F_{1,n-1}(u_1, \dots, u_r), F_{2,n-1}(u_1, \dots, u_r), \dots, F_{r,n-1}(u_1, \dots, u_r)\right)^T \,,$$

and the matrix

$$(Z_{r-1}, Z_{r-2}, \dots, Z_0) = \begin{pmatrix} {z_1}^{(r-1)} & {z_1}^{(r-2)} & \dots & {z_1}^{(0)} \\ {z_2}^{(r-1)} & {z_2}^{(r-2)} & \dots & {z_2}^{(0)} \\ \dots & \dots & \dots & \dots \\ {z_r}^{(r-1)} & {z_r}^{(r-2)} & \dots & {z_r}^{(0)} \end{pmatrix}$$

then, the solution of problem (18) can be written

$$Z_n = (Z_{r-1}, Z_{r-2}, \dots, Z_0) \cdot \mathcal{F}_{n-1}(u_1, \dots, u_r)\,. \tag{19}$$

That is, for the components:

$$\begin{aligned} z_k{}^{(n)} &= F_{1,n-1}(u_1, \dots, u_r) z_k{}^{(r-1)} + F_{2,n-1}(u_1, \dots, u_r) z_k{}^{(r-2)} + \\ &+ \dots + F_{r,n-1}(u_1, \dots, u_r) z_k{}^{(0)} = \sum_{h=1}^{r} F_{h,n-1}(u_1, \dots, u_r) z_k{}^{(r-h)}\,. \end{aligned}$$

Proof – It is well known that the solution of problem (18) is given by

$$Z(t) = \mathcal{A}^n Z_0\,. \tag{20}$$

From the results about matrix powers, it follows that

$$\begin{aligned} Z(t) = \big(&F_{1,n-1}(u_1, \dots, u_r)\mathcal{A}^{r-1} + F_{2,n-1}(u_1, \dots, u_r)\mathcal{A}^{r-2} + \\ &+ \dots + F_{r,n-1}(u_1, \dots, u_r)\mathcal{I}\big) Z_0\,. \end{aligned} \tag{21}$$

Then, taking into account the above definitions of vectors $\mathcal{F}_{n-1}(u_1, \dots, u_r)$ and $Z_1, Z_2, \dots, Z_{r-1}$ our result follows.

Remark – Note that, even if this is unrealistic, solution (21) still holds for negative values of n, assuming definition (3) for the $F_{k,n}$ functions when $n < -1$.

5 The Continuous Time Case

Theorem – Consider the Cauchy problem for the homogeneous linear differential system

$$\begin{cases} Z'(t) = \mathcal{A} \cdot Z(t) \\ Z(0) = Z_0\,, \end{cases} \tag{22}$$

where $\mathcal{A} = \mathcal{A}_{r\times r}$ is the matrix considered in Sect. 4.

Let

$$Z(t) = (Z_1(t), Z_2(t), \dots, Z_r(t))^T\,,$$

$$Z_0 = Z_0^0 = (Z_1(0), Z_2(0), \dots, Z_r(0))^T\,,$$

$$Z_0' = (Z_1'(0), Z_2'(0), \dots, Z_r'(0))^T = \mathcal{A} \cdot Z_0\,,$$

$$\dots \quad \dots \quad \dots$$

$$Z_0^{r-1} = (Z_1^{r-1}(0), Z_2^{r-1}(0), \dots, Z_r^{r-1}(0))^T = \mathcal{A} \cdot Z_0^{r-2}\,,$$

denote by $u_1 := tr(\mathcal{A})$, $u_2, \ldots, u_r := det(\mathcal{A})$ the invariants of $\mathcal{A}$, and recall again the generalized Lucas polynomials $F_{k,n}(u_1, \ldots, u_r)$, $(k = 1, 2, \ldots, r\,,\; n \geq -1)$.

Introduce the matrix $\left(Z_0^{r-1}, \ldots, Z_0', Z_0\right)$ and define the vector function

$$\mathcal{F}_{n-1}(u_1, \ldots, u_r) := \left(F_{1,n-1}(u_1, \ldots, u_r), F_{2,n-1}(u_1, \ldots, u_r), \ldots, F_{r,n-1}(u_1, \ldots, u_r)\right)^T,$$

then, the solution of problem (22) can be written

$$Z(t) = \sum_{n=0}^{\infty} \left(Z_0^{r-1}, \ldots, Z_0', Z_0\right) \cdot \mathcal{F}_{n-1}(u_1, \ldots, u_r)\, \frac{t^n}{n!}\,. \tag{23}$$

Proof – It is well known that the solution of problem (22) is given by

$$Z(t) = e^{t\mathcal{A}} Z_0\,. \tag{24}$$

From the results about matrix exponential, it follows that

$$e^{t\mathcal{A}} = \sum_{n=0}^{\infty} \frac{(t\mathcal{A})^n}{n!} = \sum_{n=0}^{\infty} \mathcal{A}^n \frac{t^n}{n!}\,, \tag{25}$$

where

$$\begin{aligned}\mathcal{A}^n = F_{1,n-1}(u_1, \ldots, u_r)\mathcal{A}^{r-1} + F_{2,n-1}(u_1, \ldots, u_r)\mathcal{A}^{r-2} + \\ + \cdots + F_{r,n-1}(u_1, \ldots, u_r)\mathcal{I}\,,\end{aligned} \tag{26}$$

so that Eq. (25) becomes

$$\begin{aligned}Z(t) = \sum_{n=0}^{\infty} \Big(F_{1,n-1}(u_1, \ldots, u_r)\mathcal{A}^{r-1} + F_{2,n-1}(u_1, \ldots, u_r)\mathcal{A}^{r-2} + \\ + \cdots + F_{r,n-1}(u_1, \ldots, u_r)\mathcal{I}\Big)\, \frac{t^n}{n!}\, Z_0\,,\end{aligned}$$

and taking into account the above positions, it follows

$$\begin{aligned}Z(t) = \sum_{n=0}^{\infty} \Big(F_{1,n-1}(u_1, \ldots, u_r)Z_0^{r-1} + F_{2,n-1}(u_1, \ldots, u_r)Z_0^{r-2} + \\ + \cdots + F_{r,n-1}(u_1, \ldots, u_r)Z_0\Big)\, \frac{t^n}{n!}\,.\end{aligned}$$

Then, Eq. (23) immediately follows by introducing the matrix $\left(Z_0^{r-1}, \ldots, Z_0', Z_0\right)$ and the vector function $\mathcal{F}_{n-1}$ defined above.

Note that the convergence of the vectorial series in any compact set K of the space $(u_1, \ldots, u_r)$ is guaranteed, as the components of $\mathcal{F}_{n-1}$ are polynomials of weight not exceeding $n-1$, and consequently are bounded in K.

5.1 *Avoiding Series Expansion by Using the Riesz–Fantappiè Formula*

By using representation (17), it is possible to avoid even the Taylor expansion in Eq. (23). In fact from the Riesz–Fantappiè formula we find:
Theorem – The solution of the Cauchy problem (22) can be found in the form

$$Z(t) = e^{t\mathcal{A}} Z_0 = \sum_{h=0}^{r-1} \left[\frac{1}{2\pi i} \sum_{j=0}^{r-h-1} (-1)^j u_j(t) \oint_\gamma \frac{e^\lambda \lambda^{r-h-j-1}}{P(\lambda, t)} \, d\lambda \right] \cdot t^h \, Z_0^h \, , \quad (27)$$

where we denoted by $u_j(t) = t^j u_j$, $(j = 0, 1, \ldots, r)$ the invariants of $t\mathcal{A}$ and by $P(\lambda, t) = \lambda^r - u_1(t)\lambda^{r-1} + \cdots + u_{r-1}(t)\lambda + u_r(t)$ its characteristic polynomial.
Proof – It is a straightforward application of the Riesz–Fantappiè formula, taking into account the definition of Z_0^h , $(h = 0, 1, \ldots, r-1)$.

6 Conclusion

We have recalled that the exponential $\exp(\mathcal{A})$ of a matrix $\mathcal{A}$ can be written as a matrix polynomial, obtained from the scalar polynomial interpolating $\exp(x)$ on the spectrum of $\mathcal{A}$, and then avoiding the Taylor expansion (12).

Furthermore, by exploiting the $F_{k,n}$ functions, and in particular the fundamental solution of a homogeneous linear recurrence relation, i.e. the generalized Lucas polynomials of the second kind, we have shown how to obtain

- The solution of discrete time evolution problems, avoiding higher powers of the matrix $\mathcal{A}$.
- The Taylor expansion of the solution of the continuous time evolution problem (22) in terms of functions of the invariants of $\mathcal{A}$, instead of powers of $\mathcal{A}$. These functions are independent of the Jordan canonical form of $\mathcal{A}$, and can be computed recursively, avoiding the knowledge of eigenvectors and principal vectors. Moreover, if the matrix is real, the $F_{k,n}$ functions are real as well, and possible complex eigenvalues does not change the form of solution.
- The Riesz–Fantappiè representation of solution of problem (22), expressed by a finite sum, and moreover avoiding even the explicit knowledge of the eigenvalues of $\mathcal{A}$, since is sufficient, using the Gerschgorin theorem, the knowledge of a circle D containing the spectrum of $\mathcal{A}$.

Therefore, the last two are, in our opinion, more convenient forms for the solution of problem (22).

References

1. M.W. Hirsch, S. Smale, R.L. Devaney, *Differential equations, Dynamical Systems and An Introduction to Chaos* (Academic Press, (Elsevier), San Diego - London, 2003)
2. E.R. Scheinerman, *Invitation to Dynamical Systems* (Dover, New York, 2012)
3. É. Lucas, *Théorie des Nombres* (Gauthier-Villars, Paris, 1891)
4. I.V.V. Raghavacharyulu, A.R. Tekumalla, Solution of the Difference Equations of Generalized Lucas Polynomials. J. Math. Phys. **13**, 321–324 (1972)
5. M. Bruschi, P.E. Ricci, An explicit formula for $f(\mathcal{A})$ and the generating function of the generalized Lucas polynomials. Siam J. Math. Anal. **13**, 162–165 (1982)
6. M. Bruschi, P.E. Ricci, I polinomi di Lucas e di Tchebycheff in più variabili, *Rend. Mat.*, S. VI, **13**, (1980), 507–530
7. P.E. Ricci, Sulle potenze di una matrice, *Rend. Mat.* (6) 9 (1976), 179–194
8. F.R. Gantmacher, *Matrix Theory* (Chelsea Pub. Co, New York, 1960)
9. R. Lidl, C. Wells, Chebyshev polynomials in several variables. J. Reine Angew. Math. **255**, 104–111 (1972)
10. R. Lidl, Tschebyscheff polynome in mehreren variabelen. J. Reine Angew. Math. **273**, 178–198 (1975)
11. T.H. Koornwinder, Orthogonal polynomials in two variables which are eigenfunctions of two algebraically independent partial differential operators, I–II. Kon. Ned. Akad. Wet. Ser. A, **77**, 46–66 (1974)
12. T.H. Koornwinder, Orthogonal polynomials in two variables which are eigenfunctions of two algebraically independent partial differential operators, III-IV. Indag. Math. **36**, 357–381 (1974)
13. K.B. Dunn, R. Lidl, Multi-dimensional generalizations of the Chebyshev polynomials, I-II. Proc. Japan Acad. **56**, 154–165 (1980)
14. R.J. Beerends, Chebyshev polynomials in several variables and the radial part of the Laplace-Beltrami operator. Trans. Am. Math. Soc. **328**(2), 779–814 (1991)

On a Geometric Model of Bodies with "Complex" Configuration and Some Movements

Ilia Tavkhelidze, Diego Caratelli, Johan Gielis, Paolo Emilio Ricci, Mamanti Rogava and Maria Transirico

Abstract Aim of this chapter is analytical representation of one wide class of geometric figures (lines, surfaces and bodies) and their complicated displacements. The accurate estimation of physical characteristics (such as volume, surface area, length, or other specific parameters) relevant to human organs is of fundamental importance in medicine. One central idea of this article is, in this respect, to provide a general methodology for the evaluation, as a function of time, of the volume and center of gravity featured by moving of one class of bodies used of describe different human organs.

Keywords Analytical representation of moving organs · Möbius strip · Torus-like bodies · Helices · Bulky links · Gaussian curvature · Center of gravity

---

I. Tavkhelidze (✉)
Department of Mathematics, Iv. Javakhishvili Tbilisi State University, University Street 2, 0186 Tbilisi, Georgia
e-mail: ilia.tavkhelidze@tsu.ge

D. Caratelli
The Antenna Company, High Tech Campus 41, Eindhoven, The Netherlands
e-mail: diego.caratelli@antennacompany.com

D. Caratelli
Tomsk Polytechnic University, 84/3 Sovetskaya Street, 634050 Tomsk, Russia

J. Gielis
Department of Bioengineering Sciences, University of Antwerp, Wilrijk, Antwerpen, Belgium
e-mail: johan.gielis@uantwerpen.be

P.E. Ricci
Department of Mathematics, International Telematic University Uninettuno, Roma, Italy
e-mail: paoloemilioricci@gmail.com

M. Rogava
Medical Clinic "Janmrtelobis Centri", Tbilisi, Georgia
e-mail: mamantirogava@mail.ru

M. Transirico
University of Salerno, Via Ponte don Melillo, 84084 Fisciano, Italy
e-mail: mtransirico@unisa.it

J. Gielis et al. (eds.), *Modeling in Mathematics*,
Atlantis Transactions in Geometry 2, DOI 10.2991/978-94-6239-261-8_10

2000 Mathematics Subject Classification 53A05 · 51B10

1 Introduction

Recently, the mathematical modeling of human organs is attracting a great deal of interest in modern sciences. As an example, the imaging of heart is essential to collect kinematic and anatomical information in the form of structural boundaries which are afterwards used in suitable computational fluid dynamics models for the evaluation of the blood motion flow within the organ. This in turn is important to analyze heart and valve biodynamics and the mechanisms ensuring their proper functioning. The accurate volumetric description of human organs is necessary to assess the relevant, it's easy to figure out that the accurate volumetric description of human organs is beneficial for aiding the assessment of the relevant functional behavior and early detection of possible pathologies. In this context, organs can be conveniently modeled in terms of generalized twisted and rotated bodies, whose analytical or algorithmic representations are thoroughly discussed in [1–4] and for education process in [5]. Here emphasis is put on the evaluation of the geometrical characteristics of these bodies and, in particular, the relevant volume and center of gravity as functions of time due to arbitrary motion. Here it is also worth noting that the displacement of the mentioned bodies can be described analytically in terms of elementary displacements [1, 6]. The first attempt at an analytical representation of a concrete *GML* surface of this class of geometric shapes in statics was implemented in the year 2000 in article [7]. Further in 2006 [1], a very wide class of geometric figures were described which was called - **Generalized Twisting and Rotated figures**, sometimes called surfaces of Revolution see e.g. [2]. The geometric meaning of functions and parameters included in the analytic representation were identified. Also, in parallel, step by step, the Elementary Displacements of these figures were described. Since 2011 [6, 8, 9] it became possible to analytically describe the figures and their movements whose radial cross-section and the basic lines are Gielis curves. In this part we offer a description of a subclass of figures whose basic lines are three-dimensional toroidal lines. Also, we show the possibility of an analytic representation or separation of the wide class of complicated movements into the elementary displacements. The idea of the methodology of realization of the present analytic representation originated from Gaspard Monge and is based on the idea of representing a complex machine as a superposition of elementary mechanisms.

1.1 Notations

In this article we use following notations:

- X; Y; Z denote, as usual, the Cartesian coordinates;
- t - denote time -$t \in [0, \infty)$;
- τ, ψ, θ - are local space coordinates or parameters in parallelogram;

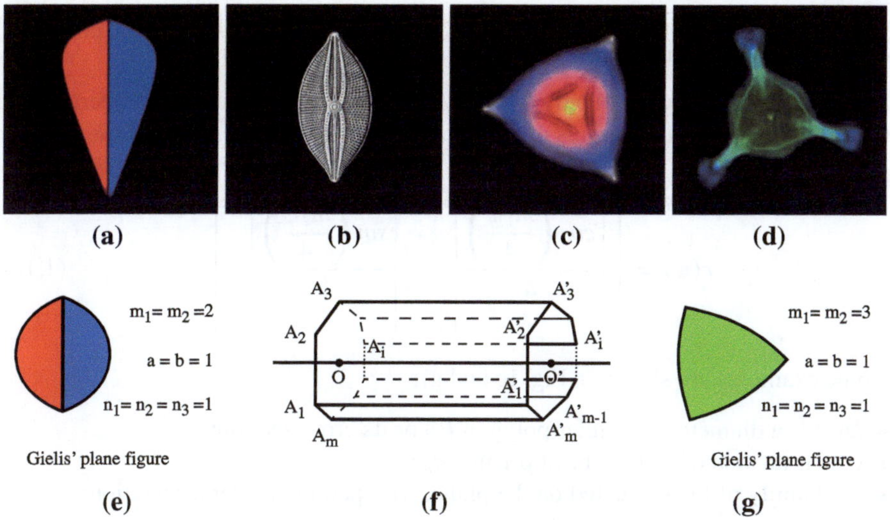

Fig. 1 Examples of natural and pure geometric plane figures P_m - with different parameters of symmetry; Case (h) is an orthogonal prism PR_m

1. $\tau \in [\tau_*, \tau^*]$, with $\tau_* \le \tau^*$, where $\tau_* \le \tau^*$ usually are non-negative constants;
2. $\psi \in [0, 2\pi]$;
3. $\theta \in [0, 2\pi h]$, with $h \in \mathbb{R}$ (Real number).

(1.1.1)

- $PR_m \equiv A_0A_1 \ldots A_{m-1}A'_0A'_1 \ldots A'_{m-1}$ denotes an orthogonal prism, whose ends $A_0A_1 \ldots A_{m-1}$ and $A'_0A'_1 \ldots A'_{m-1}$ are P_m and P'_m;
- $P_m \equiv A_0A_1 \ldots A_{m-1}$ - is an "**Plane figure with m-symmetry**", in particular P_m is a regular polygon and m is the number of its angles or vertices. In the general case the edges of regular polygons are not always straight lines (A_iA_{i+1} may be, for example: the edge of epicycloid, or one edge of hypocycloid, or part of lemniscate of Bernoulli, and so on); Correspondingly: PR_0 - is a segment and P_0 - point; P_1 - plane figure without symmetry; PR_∞ - are orthogonal cylinders whose ends respectively are and P_∞ - circle; (some examples are shown in Fig. 1: case a. is a P_1; cases b. and e. are Natural and Pure Geometric objects P_2, and cases c., d. and g. are P_3);

Let

$$x = r_*(\tau, \psi, \theta, t); \quad z = r^*(\tau, \psi, \theta, t); \tag{1.1.2}$$

or

$$x = r(\tau, \psi, \theta, t)cos\psi; \quad z = r(\tau, \psi, \theta, t)sin\psi; \tag{1.1.2*}$$

be analytic representations of the regular polygon $P_m(t)$ (more precisely the shape of figures in the moment t), usually $r_*(0, 0, t) = r^*(0, 0, t) = 0$ and let point $(0, 0, t)$ be a center of symmetry of this polygon. Formula (1.1.2) or (1.1.2*) gives us the opportunity to describe the case when the shape of polygon depends on the position of cross section (θ) and can continuously change over time (t).

Remark 1.1 A wide class of polygons may be represented by the Gielis Superformula (1.1.3) with different meanings of parameters: a, b, m_1, m_2, n_1, n_2, and n_3

$$r(\psi) = \left[\left| \frac{cos\left(\frac{m_1\psi}{4}\right)}{a} \right|^{n_2} + \left| \frac{sin\left(\frac{m_2\psi}{4}\right)}{b} \right|^{n_3} \right]^{-\frac{1}{n_1}} \tag{1.1.3}$$

(some examples are shown in Fig. 1e and g);

- $D(r)$ is a diameter of regular polygon P_m or its cross section;
- OO' is an axis of symmetry of prism PR_m;
- L_ϱ- Family of lines situated on the plane, with parametric representation:

$$L_\varrho = \{X = R_*(\varrho, \theta, t);\; Y = R^*(\varrho, \theta, t) :\; where\ \varrho \in [0, \infty);\; \theta \in [0, 2\pi h],\ h \in \mathbb{Z}\} \tag{1.1.4}$$

or particularly

$$L_\varrho = \{X = R(\varrho, \theta, t)cos\theta;\; Y = R(\varrho, \theta, t)sin\theta\} \tag{1.1.4*}$$

We assume the following hypotheses for each fixed $t \in [0, \infty)$:

(i) For any $\varrho_1, \varrho_2 \in [0, \varrho^*]$, $\varrho_1 \neq \varrho_2$, corresponding lines L_{ϱ_1} and L_{ϱ_2} have no common points.

(ii) If L_ϱ is a closed curve, then for every fixed $\varrho \in [0, \varrho^*]$ the corresponding R_* and R^* functions are 2π-periodic of argument θ;

Remark 1.2 A wide class of L_ϱ curves family may be represented by Gielis Superformula (1.1.3) with different meanings of parameters: a, b, m_1, m_2, n_1, n_2, and n_3

- *Let* $g(\theta) : [0, 2\pi] \to [0, 2\pi]$ be an arbitrary function and for every $\Theta \in [0, 2\pi]$ exist $\theta \in [0, 2\pi]$, such that
$$\Theta = g(\theta) \tag{1.1.5}$$
- $mod_m(n)$ denotes a natural number smaller than m such that, for arbitrary numbers $m \in \mathbb{N}$ and $n \in \mathbb{Z}$ a unique representation exists for $n = km + j = km + mod_m(n)$, where $k \in \mathbb{Z}$ and $j \equiv mod_m(n) \in \mathbb{N} \bigcup \{0\}$ the sets of the natural and integer numbers should be denoted using the double-struck capital $\mathbb{N}$ and $\mathbb{Z}$ symbols;
- $$\mu\ \ or\ \ \nu \equiv \begin{cases} \dfrac{n}{m} & m \in \mathbb{N}\ and\ \ n \in \mathbb{Z} \\ n\ \ when & m = \infty\ \ and\ \ n \in \mathbb{Z} \end{cases} \tag{1.1.6}$$

1.2 Analytic Representation of Generalised Twisted and Rotated Figures and Generalised Möbius-Listing's Bodies

Definition 1.1 Generalized Twisting and Rotating figures $GTR_m^n\{\mu\}$ - are obtained if the axis OO' of the prism PR_m is transformed in a spatial (or plane) line (basic line) with characteristic μ (see in (1.1.6)) and at the same time the prism is twisted with respect to this line.

According to this definition and notations the class of $GTR_m^n\{\mu\}$ figures has been described analytically in one of the following formulas (see in [1] or [9])

$$\begin{cases} X(\tau,\psi,\theta) = R_*\left(\left[R + r_*(\tau,\psi)cos(\mu g(\theta)) - r^*(\tau,\psi)sin(\mu g(\theta))\right],\theta\right) \\ Y(\tau,\psi,\theta) = R^*\left(\left[R + r_*(\tau,\psi)cos(\mu g(\theta)) - r^*(\tau,\psi)sin(\mu g(\theta))\right],\theta\right) \\ Z(\tau,\psi,\theta) = K(\theta) + r_*(\tau,\psi)sin(\mu g(\theta)) + r^*(\tau,\psi)sin(\mu g(\theta)) \end{cases} \quad (1.2.1)$$

or

$$\begin{aligned} X(\tau,\psi,\theta) &= [R(\theta) + r(\tau,\psi,\theta)cos(\psi + \mu g(\theta))]\,cos\theta \\ Y(\tau,\psi,\theta) &= [R(\theta) + r(\tau,\psi,\theta)cos(\psi + \mu g(\theta))]\,sin\theta \\ Z(\tau,\psi,\theta) &= K(\theta) + r(\tau,\psi,\theta)sin(\psi + \mu g(\theta)), \end{aligned} \quad (1.2.1^*)$$

where respectively:

- arguments τ, ψ, θ are defined in (1.1.1); Additionally, we note that the number h in (1.1.1) is the number of coils of the $GTR_m^n\{\mu\}$ bodies if this number is natural, but in the general case h is arbitrary real number (in this case geometric meaning of h is "polar length" of corresponding body);
- functions R_* and R^* or R describe the **shape of plane basic line**, defined correspondingly by (1.1.4) or (1.1.4*);
- functions r_* and r^* or $r(\tau, \psi, \theta)$ describe the **shape of radial cross section**, defined correspondingly by (1.1.2) or (1.1.2*);
- function $g(\theta)$ describes **the rule of twisting of the cross section around the basic line** and it is defined by (1.1.5);
- number μ describes the **characteristic of twisting** and it is defined by (1.1.6);
- function $K(\theta)$ describes the **law of vertical stretch of figure** and it may be an arbitrary smooth function;
- number R in formula (1.1.2) describes the **radius** and it may be some fixed real nonnegative number.

These figures can be described in many ways (see [2, 3, 10]), but the difference with our proposed analytical representation is that each function and each parameter in the formula has a geometric meaning. So you can create targeted pre-conceived figures, as well as determine with sufficient accuracy the function or parameters of the existing $GTR_m^n\{\mu\}$ figures!

First in 2006 [1], this formula has given us the opportunity of an analytic representation of a wide class of figures. But starting in 2010 (see e.g. [3, 9]), using the Gielis Superformula twice - once for the description of shape of plane basic line and the second time for the description of shape of radial cross section - the range of figures (shapes) significantly expanded. It gives us the opportunity not only to describe the class of figures but also to study some geometric properties and explore new phenomena in some classic and unusual figures. Detailed figures have been described in the article [1], but here we give some examples:

Case A. **Helix Bodies** In this case the functions in expressions (1.2.1) or (1.2.1*) have the following restrictions:
∘ Basic line L_ϱ is some closed line (1.1.4*) or in representation (1.1.4) functions R_* and R^* satisfy the condition ii.);
∘ Law of vertical stretch of figure $K(\theta) \neq 0$;
∘ Number m in P_m does not depend on μ, see defining expression (1.1.6);
∘ Number h in (1.1.1) is a number of complete coils of helix bodies if $h \in \mathbb{N}$, (but in general case h is arbitrary real number);

General Remarks:
⋆ If the function $K(\theta) = K \cdot \theta \neq 0$, then (1.1.2) or (1.1.2*) define a set of **helix bodies** with constant step of vertical stretch. The number K is a measure of this step;
⋆ If $K(\theta) = K \cdot \theta \neq 0$, and number K is greater than the diameter of the regular polygon P_m divided by 2π (In particular, in case $|K| > \max(r(\theta)/\pi)$, then functions in the representations (1.2.1) or (1.2.1*) define a set of **helix bodies** without self-intersection points;

Example A. a. If functions (1.1.2) or (1.1.2*), define the shape of the radial cross section of helix bodies have the following form:

$$x = r_*(\tau, \psi, t) \equiv \tau; \quad z = r^*(\tau, \psi, t) = sin\tau; \quad g(\theta) \equiv \theta; \quad h = 4;$$
$$L_R \textit{ is a circle with radius } R > \tau^*; \quad K(\theta) = K \cdot \theta \neq 0.$$

a typical rendering of (1.2.1) is then shown in Fig. 2a;

Example A. b. If the basic line L_R is a rectangle i.e. in formula (1.1.4*)

$$R(\varrho, \theta, t) = \frac{\varrho}{|cos\theta| + |sin\theta|} \tag{1.2.2}$$

$g(\theta) \equiv \theta$ and function in (1.1.2*) is a parametric representation of epicycloid, hypocycloid or ellipse, then (1.2.1*) define a set of helix bodies with rectangular basic line and with corresponding radial cross section. An example is shown in Fig. 2b;

Example A. c. In Fig. 2c, a helix body with radial cross section - lemniscate and whose basic line L_R is a cross section of a *Nuphar luteum* (Gielis Superformula (1.1.3) when $a = b = 1, m_1 = m_2 = 3, n_1 = 4.5$ and $n_2 = n_3 = 10$) is shown;

Example A. d. Some different examples of helix bodies whose basic lines are circles or radial cross sections are epicycloid, hypocycloid are shown in Fig. 3a and b;

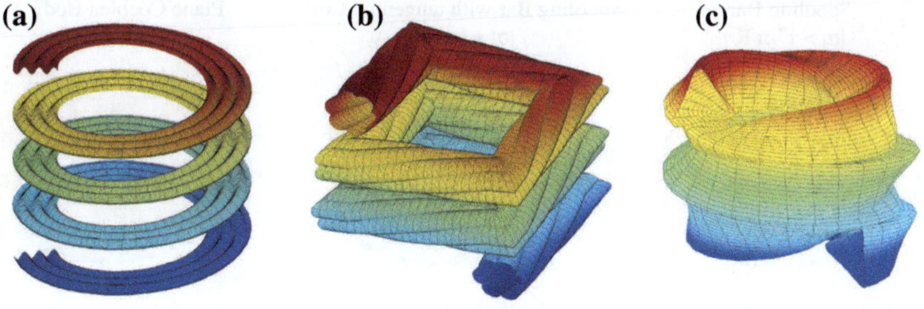

Fig. 2 Examples of the helix bodies which have different basic lines and radial cross-sections: **a** circle and sinusoid, **b** rectangle and epicycloid, **c** Circle and Nuphar luteum

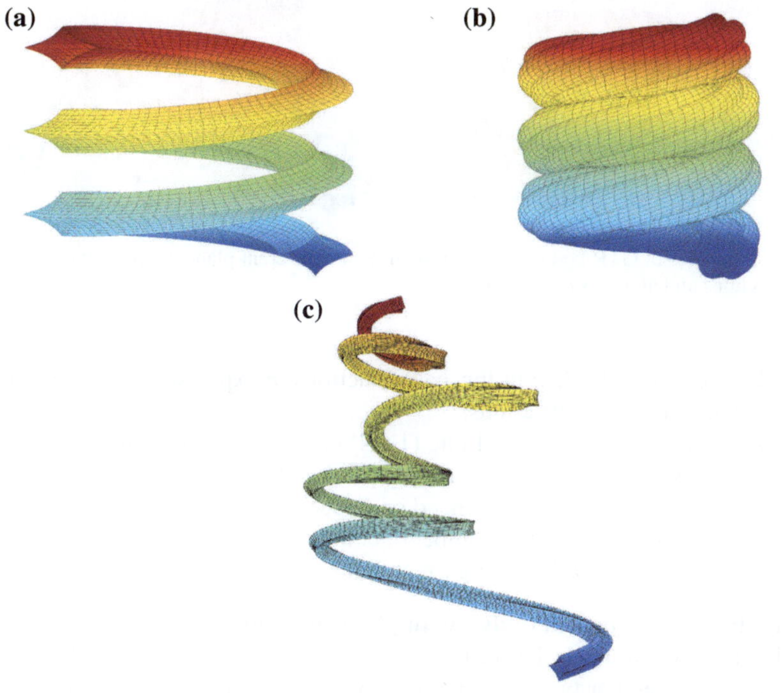

Fig. 3 Examples of the GTR bodies which have different basic lines, radial cross-sections and numbers of twisting

Example A. e. If the basic line L_R (1.1.4) and the radial cross section (1.1.2) are given on Gielis Equisetum, i.e. in (1.1.3) correspondingly $a = b = 1$; $m_1 = m_2 = 7$, $n_1 = 10$, $n_2 = n_3 = 6$ the law of vertical stretch of figure is $K = 2$ and in (1.1.1) $\theta \in [0, 6\pi]$, then the graphic image of corresponding figure according to the (1.2.1*) is given in Fig. 3c.

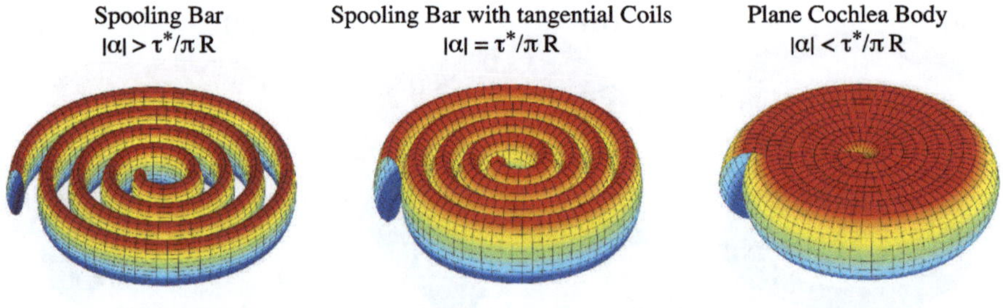

Fig. 4 Examples of the GTR bodies whose basic line is a plane spiral and radial cross sections are circles with different radii

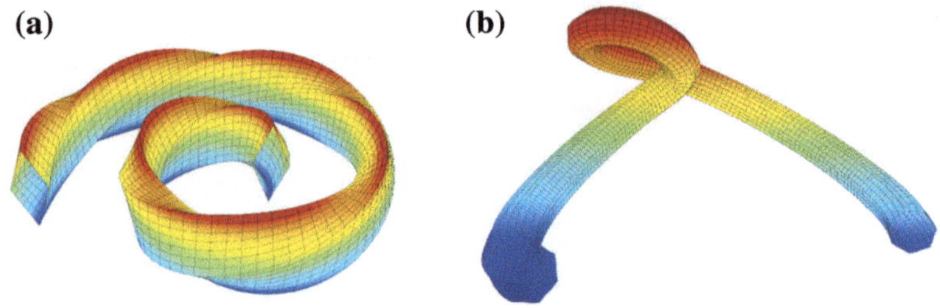

Fig. 5 Examples of GTR bodies whose basic lines are different plane or space spiral lines, radial cross sections and numbers of twisting

Case B. **Cochlea Bodies** In this case functions in expressions (1.2.1) or (1.2.1*) have the following restrictions:

∘ Basic line L_ϱ is some spiral line (1.1.4*) or in representation (1.1.4) functions R_* and R^* satisfy the condition i.) except when $\theta = 0$;

∘ Number m in P_m does not depend on μ, see defining expression (1.1.6);

∘ Number h in (1.1.1) is the number of complete coils of cochlea bodies if $h \in \mathbb{N}$, (but in general case h is arbitrary real number);

Example B. a. **Cochlea body - with plane basic line**. $K(\theta) = 0$. If basic line L_R (1.1.4) is a spiral line i.e. in (1.1.4*) $R(\theta) = R(1 + \alpha \cdot \theta)$; in (1.1.5) $g(\theta) = \theta$ (this means regular twisting) and $m = \infty$ (this means P_∞ is circle); h in (1.1.1) is a number of coils, then the representation (1.2.1*) defines the following set of figures (see Fig. 4):

⋆ **Spooling Bar** - if $|\alpha| > \tau^*/\pi R$ (example Fig. 4a);

⋆ **Spooling Bar with tangential coils** - if $|\alpha| = \tau^*/\pi R$ (example Fig. 4b);

⋆ **Plane Cochlea body** - if $|\alpha| < \tau^*/\pi R$ (example Fig. 4c);

The example of the spooling bar with Radial cross section - rectangle (i.e. in (1.1.2*) function r for example, may be represented by (1.2.2)), with regular twisting is presented in Fig. 5a; For a visual comparison, we show the example of the $GTR^n_m\{\mu\}$ bodies with radial cross section Gielis Equisetum (see example c. in Fig. 3 but now $\theta \in [0, 2\pi]$ and K = 0 is given in Fig. 5b.

Example B. b. Cochlea body - with spatial basic line. $K(\theta) \neq 0$ Functions in (1.2.1) or (1.2.1*) have the following geometric meaning and so have restrictions:

o Functions in (1.1.4) (i.e. plane basic line L_ϱ) and $K(\theta)$ (i.e. law of vertical stretch) describe real space basic line of cochlea body;
o Functions in (1.1.2) or (1.1.2*) describe the shapes of the radial cross section of cochlea body;
o Number h from (1.1.1), more precisely $[h]$ - integer part of h, defines the number of complete coils of cochlea body;
o Number m in P_m does not depend on μ, see defining expression (1.1.6);
o Function $g(\theta)$ in (1.1.5) defines the law of twisting of coils around basic line of cochlea body;

Some Examples of cochlea bodies with different configurations are presented in Fig. 6.

Case B. Generalised Möbius-Listing's bodies and Toroidal figures
Formula (1.2.1) give us possibility of the analytic representation of this very important sub-class of $GTR^n_m\{\mu\}$ figures.

Definition 1.2 Generalised Möbius-Listing's bodies $GTR^n_m\{\mu\}$ - are obtained by identifying of the opposite ends of the prism PR_m in such a way that:

(A) For any integer $n \in \mathbb{Z}$ and $i = 0, 1, \ldots, m-1$ each vertex A_i coincides with $A'_{i+n} \equiv A'_{mod_m(i+n)}$, and each edge A_iA_i+1 coincides correspondingly with the edge $A'_{i+n}A'_{i+n+1} \equiv A'_{mod_m(i+n)}A'_{mod_m(i+n+1)}$.
(B) The integer $n \in \mathbb{Z}$ is the number of rotations of the end of the prism with respect to the axis OO' before the identification; If $n > 0$ the rotations are counter-clockwise, and if $n < 0$ then rotations are clockwise.
(C) Axis OO' after identifying is transformed in a closed space (or plane) line (basic line) with characteristic ν from (1.1.6).

According to Definition 1.2 the axis of symmetry is required to be transformed into a closed plane or spatial line.

1. In this part, without loss of generality, we shall sometimes use the analytic representation (1.2.3), which is a one of particular variant of formula (1.2.1) or (1.2.1*);
2. We always assume that rotation of the ends of the prism is semi-regular along the middle line OO' (this means $g(\theta) = \nu \cdot \theta$). So the analytic representation (1.2.1) can be rewritten in following form:

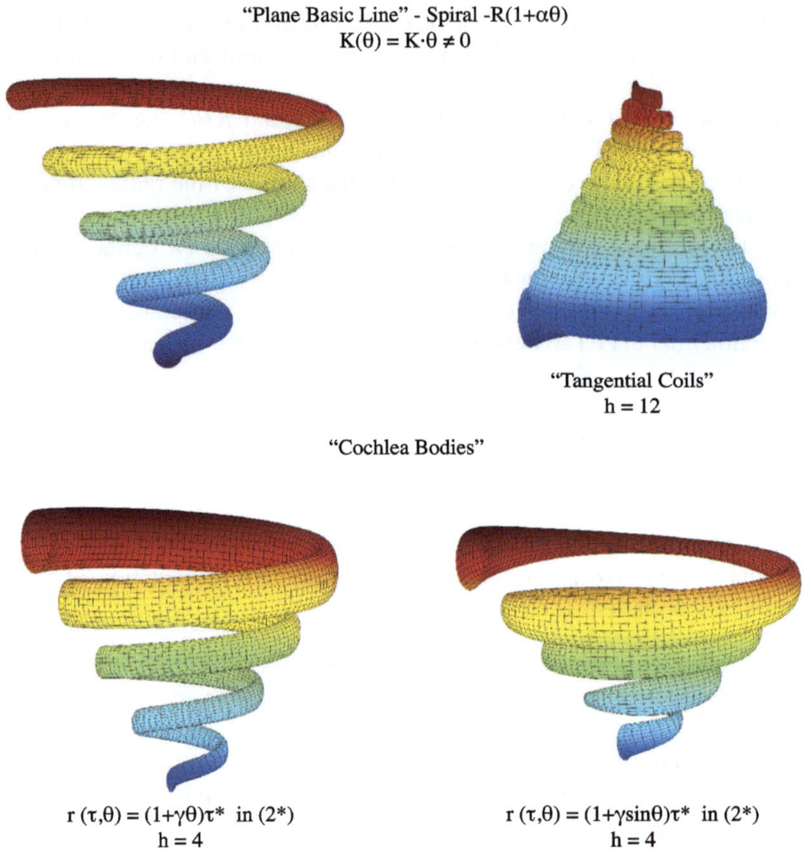

Fig. 6 Examples of GTR bodies whose basic lines are different space spiral lines, radial cross sections are variables and numbers of twisting are different

$$
\begin{aligned}
X(\tau, \psi, \theta) &= \left[R + \widetilde{R}cos\left(\frac{\widetilde{n}\theta}{\widetilde{m}}\right) + r(\tau, \psi)cos\left(\psi + \frac{n\theta}{m}\right)\right] cos\theta \\
Y(\tau, \psi, \theta) &= \left[R + \widetilde{R}cos\left(\frac{\widetilde{n}\theta}{\widetilde{m}}\right) + r(\tau, \psi)cos\left(\psi + \frac{n\theta}{m}\right)\right] sin\theta \\
Z(\tau, \psi, \theta) &= \widetilde{R}sin\left(\frac{\widetilde{n}\theta}{\widetilde{m}}\right) + r(\tau, \psi)sin\left(\psi + \frac{n\theta}{m}\right).
\end{aligned} \tag{1.2.3}
$$

⋆ if in (1.2.3) the function $r(\tau, \psi) \equiv 0$, then the corresponding line defines the shape of the spatial basic line with characteristic $\widetilde{n}/\widetilde{m}$;

Remark 1.3 This analytic representation (1.2.3) when $\widetilde{R} \geq 0$ and $r(\tau, \psi) \equiv 0$ describes **torus lines** (i.e. lines lying on the surface of a torus) and:

a. if $\nu \equiv \dfrac{\widetilde{n}}{\widetilde{m}} \in \mathbb{Z}$ (integer number), then corresponding line make $\widetilde{n}$ coils after one rotation around the torus (see e.g. Fig. 7. $\nu = 4$ and 20);

b. if $\nu \equiv \frac{\widetilde{n}}{\widetilde{m}} \in \mathbb{Q}$ (rational number), then corresponding line make $\widetilde{n}$ coils after $\widetilde{m}$ rotations around the torus (see e.g. Fig. 7. $\nu = \frac{1}{5}, -\frac{1}{5}, \frac{2}{5}, \frac{3}{5}, \frac{4}{5}, \frac{21}{5}, \frac{109}{5}, \frac{151}{5}$);

c. if $\nu \equiv \frac{\widetilde{n}}{\widetilde{m}} \in \mathbb{R}\backslash\mathbb{Q}$ (irrational number), then this line make infinite coils after infinite rotations around the torus without self-intersections;

⋆ Each meaning of the parameter defines corresponding element of the fundamental group of torus [11, 12];

Remark 1.4 The main difference between representations (1.2.1) and (1.2.3) is - the first one gives us the opportunity to represent *GTR* and *GML* figures with plane basic line; more precisely functions R_* and R^* defined according to (1.2.1) describe **plane shape of basic line** and function $K(\theta)$ defines the behavior of basic line in space (if function $K(\theta)$ is a 2π -periodic function, then the figure has spatial the basic line, but if $K = const.$, then the basic line of figure is a classic helix).

On the other hand using (1.2.3) it is more convenient to represent corresponding figures with wide class of the space basic lines, for example with torus line - in this case the line is defined by characteristic ν and only if $\nu = 0$ the corresponding figure has a plane basic line.

⋆ Both analytic representations are based on the same ideology, but have their own specific properties and are able to describe the different classes of *GTR* and *GML* figures, complementing each other.

It should be noted, that in (1.2.3):

- Function, which defined in (1.1.5), is always $g(\theta) = \theta$;
- The number $\mu \equiv \frac{n}{m}$ describes the **characteristic of twisting** and it is defined by (1.1.6), where m is the number of angles or vertexes of the regular polygon P_m, (shape of the radial cross section) and n is a number of twisting of the cross section of the prism before identification of their ends.

Definition 1.3 **Basic line** of the $GML_m^n\{\nu\}$ body for each $\nu \in \mathbb{Q}$, is a continuous closed, in the general case, spatial line on which the axis of symmetry OO' of the prism transforms after identifying the ends of the prism.

Definition 1.4 **Rib** of the $GML_m^n\{\nu\}$ body for each $\nu \in \mathbb{Q}$, is a continuous closed line, in which are situated only the vertices of the radial cross sections (plane figures) of this body (i.e. torus line with characteristic ν,).

Definition 1.5 **Side** of the $GML_m^n\{\nu\}$ body for each $\nu \in \mathbb{Q}$, is a continuous closed surface, in which only the sides of the radial cross section (plane figures) of this body are situated.

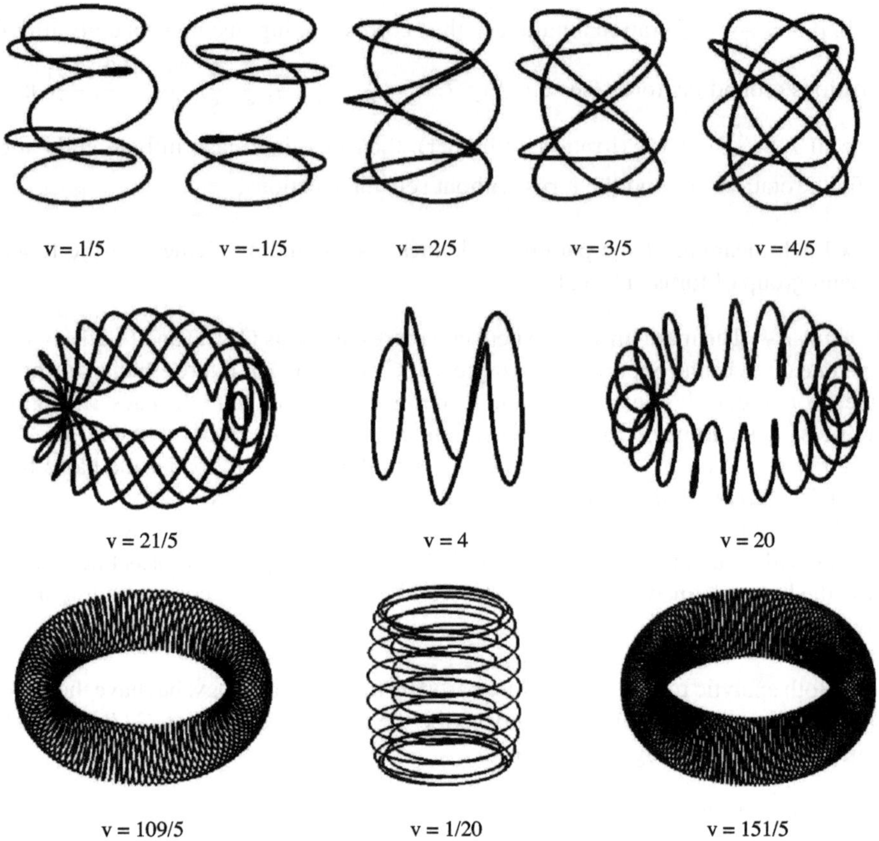

Fig. 7 Examples of torus lines with different characteristics

Notation $GML_m^n\{\nu\}$ - is Generalized Möbius Listing's body whose basic lines are **torus lines** (some examples of spatial basic lines (i.e. torus lines) are presented in Fig. 7), with characteristic $\nu = \frac{n}{m}$;

$\star$ Correspondingly $GML_m^n\{0\} \equiv GML_m^n$ and this means that basic line always is a circle!

Example C. a. **Different variants of** $GML_2^n\{0\}$ or any figures can be made from paper strips, with some assumptions on its deformability. According to the analytic representation (1.2.3) for different meanings of functions we have the possibility to describe some classic geometric figures. In particular, if in (1.2.3) $\widetilde{R} \equiv 0$, $r(\tau, \psi) \equiv \tau$, $m = 2$:

$\star$ and $n = 1$, then $GML_2^1\{0\}$ becomes a classical (or regular) Möbius band (or strip) (see e.g. [1] and Fig. 8h);

$\star$ $n = 0$ and (1.2.3) is a one to one representation then, as defined, corresponding graphic or $GML_2^0\{0\}$ figure may be regular surfaces: cylinder, ring, or frustum

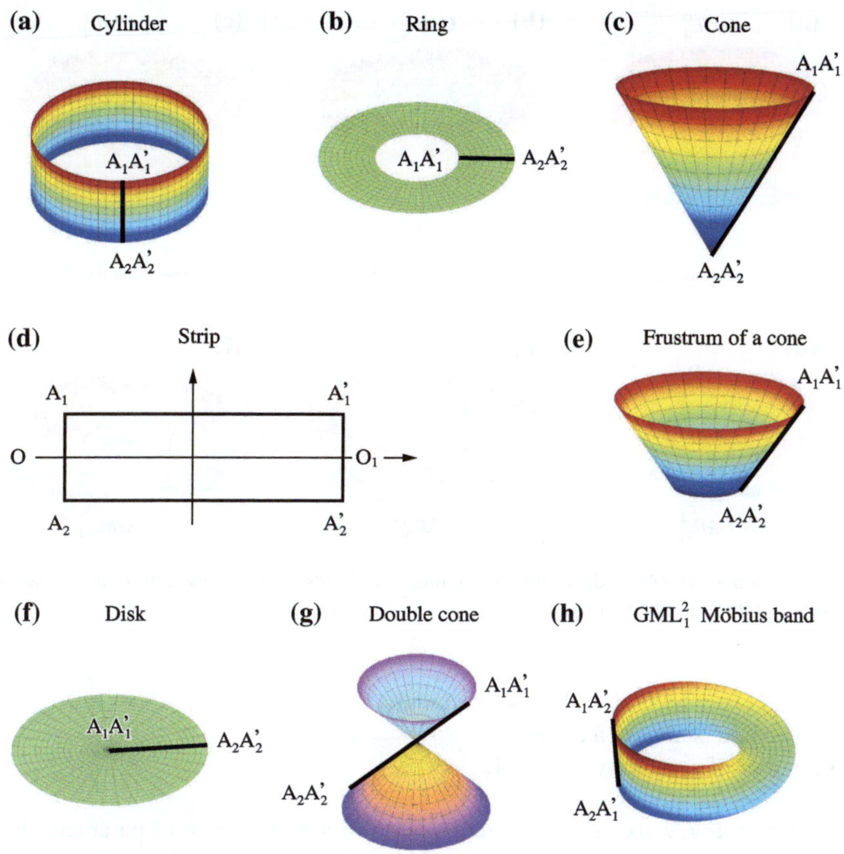

Fig. 8 Examples of GML surfaces

cone (see e.g. in Fig. 8a, b and e);
⋆ $n = 0$ and (1.2.3) is not one to one relation then, corresponding $GML_2^0\{0\}$ figures may be "irregular surfaces" (Jacobian determinants of the expression (1.2.3), which describe these surfaces have points of degeneracy): disk, cone, or double cone (see e.g. in Fig. 8f, c and g);
⋆ Always in these cases the axis OO' after identifying transfers in a circle, i.e. $\nu \equiv 0$.

Example C. b. **Different variants of** $GML_4^n\{0\}$ can be made from quadrangular prisms, with some assumptions on its deformability.

According to the analytic representation (1.2.3) for different meaning of functions we have the possibility to describe some classic and non-classic geometric figures. All of these figures have three common characteristics:

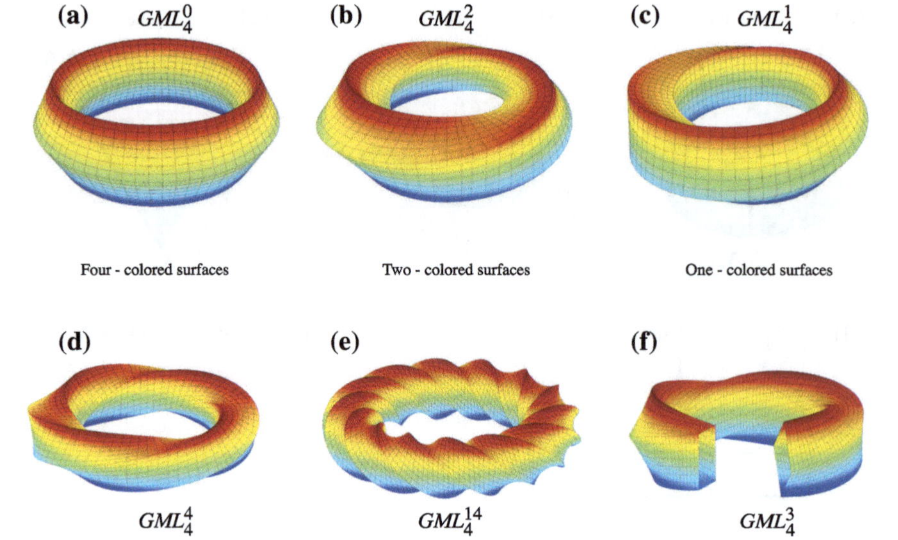

Fig. 9 Examples of GML bodies whose basic lines are circles, radial cross sections are rectangles, and have different numbers of twisting

⋆ radial cross section is always a rectangle;
⋆ basic line is always a circle;
⋆ number of symmetry is $m = 4$;

Above, in Fig. 9 six different figures with different meaning of parameter n are shown $GML_4^0\{0\}$, $GML_4^1\{0\}$, $GML_4^2\{0\}$, $GML_4^3\{0\}$, $GML_4^4\{0\}$, $GML_4^{14}\{0\}$.

Remark 1.5 **General remarks about this analytic representation and its geometric meaning**:

⋆ If in (1.1.1) $\tau_* = 0$ and $\tau^* > 0$, then all analytic representations (1.2.1), (1.2.1*), (1.2.3) describe a real three-dimensional geometric object or **Solid Body**;

⋆ If in (1.1.1) $\tau_* > 0$ and $\tau_* \neq \tau^*$, then all analytic representations (1.2.1), (1.2.1*), (1.2.3) describe a real three-dimensional geometric object similar to a **pipe**, with thickness $\tau^* - \tau_*$, but if this difference is very small (i.e. $\tau^* - \tau_* << 1$) we have a **Shell**;

⋆ If in (1.1.1) $\tau = const.$, then all analytic representations (1.2.1), (1.2.1*), (1.2.3) describe corresponding **surface**;

⋆ If in (1.1.1) $\tau = const.$ and at the same time argument $\psi = const.$, then all analytic representations (1.2.1), (1.2.1*), (1.2.3) describe a **line**, which is situated on the corresponding surface.

Remark 1.6 **General remarks about the analytic representation of generalized Möbius-Listing's bodies** Always, all analytic representations (1.2.1), (1.2.1*), (1.2.3) describe:

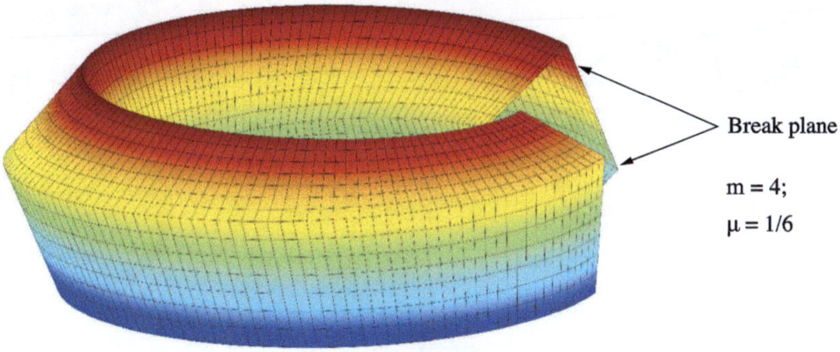

Fig. 10 An example of GML body with break along its radial cross section

⋆ 1. Toroidal body with **break** along the radial cross section, if m (number of vertexes of the regular polygon P_m) is finite, but it is different from denominator of μ from (1.1.6) (see. for example Fig. 10);

⋆ 2. Generalized Möbius-Listing's body with **j-colored surface**, or more precisely $GML_m^n\{\mu\}$ body with j different sides (see Definition 1.5), if j is a greatest common divisor of numbers m and $mod_m(n)$. This means - it is possible to paint the surface of this figure in j different colors without taking away of the brush. It is prohibited to cross the edge of this figure;

• When $m = 2$, then 1 or 2 - colored surfaces are identical to 1 or 2 - sided surfaces respectively;

⋆ 3. Generalized Möbius-Listing's body with **m-colored surface**, if $j = 0$. This means, either the number of torsion n is divisible by the number of symmetry m, or it is zero;

⋆ 4. Generalized Möbius-Listing's body with **one rib**, if $j = 1$ (see Definition 1.4). This means the boundary edge of body is one space line;

⋆ 5. Toroidal body with circular radial cross section, if $m = \infty$;

Some different examples of $GML_m^n\{\mu\}$ bodies are presented in Fig. 11.

The shape of the cross sections and the shape of the basic lines of these bodies are given by Gielis superformula (1.1.3) with different meanings of parameters.

1.3 Elementary Displacements

For engineers, the result attributed to Gaspard Monge is well known - a **complex machine is a superposition of elementary mechanisms**. We took advantage of this result, as the basic idea and suggested the possibility of **presenting a complex movement by the superposition of elementary displacements**. With simple geometric conditions we were able to describe the analytical rather wide class of object movements.The basis has been the analytical representation (1.2.1), and this

Fig. 11 Examples of GML bodies with different: basic lines and radial cross sections, which are described by the Gielis Superformula (1.1.3)

allowed to study the geometric behavior of curves and surfaces defined by different functions and parameters that are not dependent on the time of the argument.

In this part of the chapter we give parametric representation of a wide set of motions of geometric figures, surfaces, lines and points under following restrictions:

∘ 1. OO' - axis of symmetry (middle line) of the prism PR_m transforms into some line L_ϱ, i.e. basic line;

∘ 2. Rotation of the end of the prism is semi-regular along the middle line OO'. This means the function $g(\theta)$ is known;

∘ 3. Rotation of the end of the prism is semi-regular along the central line in time. This means the function $g(\theta, t)$ depends on the time argument t and this regularity is known;

∘ 4. Rotation of the end of the cross section of prism around of basic line semi-regular in time - **Rotation in orbit**. This means the function in (1.1.2) depends on time argument.

Under these restrictions the analytic representation of this class is given by the following formula

$$
\begin{aligned}
&X(\tau, \psi, \theta, t) = T_1(t) + \\
&R_* \left([R(\theta, t) + r_*(\tau, \psi, \theta, t) \cos(\psi + \mu g(\theta) + \xi\omega(t)) \right. \\
&\quad \left. -r^*(\tau, \psi, \theta, t) \sin(\psi + \mu g(\theta) + \xi\omega(t))] \,, \theta + \varsigma(t), t \right) \\
&Y(\tau, \psi, \theta, t) = T_2(t) + \\
&R^* \left([R(\theta, t) + r_*(\tau, \psi, \theta, t) \cos(\psi + \mu g(\theta) + \xi\omega(t)) \right. \\
&\quad \left. -r^*(\tau, \psi, \theta, t) \sin(\psi + \mu g(\theta) + \xi\omega(t))] \,, \theta + \varsigma(t), t \right) \\
&Z(\tau, \psi, \theta, t) = T_3(t) + K(\theta, t) + \\
&r_*(\tau, \psi, \theta, t) \sin(\psi + \mu g(\theta) + \xi\omega(t)) - r^*(\tau, \psi, \theta, t) \cos(\psi + \mu g(\theta) + \xi\omega(t))
\end{aligned}
\tag{1.3.1}
$$

where:

- arguments τ, ψ, θ, t are defined by (1.1.1);
- $T_i(t)$ for any $i = 1, 2, 3$ are arbitrary smooth functions and they describe the movement of a geometric object, as a unit, in space;
- functions R_* and R^* describe the **orbit** or the shape of plane basic line and are defined correspondingly by (1.1.4) or (1.1.4*);
- functions r_* and r^* describe the **shape of the radial cross section** corresponding to (1.1.4) or (1.1.2*); Usually in previous part of this chapter these functions do not depend on the time argument, but now they strictly depend on the time argument t. This means that the shape of radial cross section can vary in time;
- the function $g(\theta)$ describes the rule of **twisting** of the cross section around the basic line and it is defined by (1.1.5) and it characterizes the object statically;
- $\omega(t)$ may be arbitrary real smooth functions and they describe the rule of **twisting** of the cross section around the basic line **in time**;
- $\varsigma(t)$ may be arbitrary real smooth functions and they describe the rule of **rotation** of this geometric object around the axis OZ in time;
- $K(\theta, t)$ may be arbitrary real smooth functions and they describe the law of **vertical stretch** of the figure in time and along the basic line;
- numbers μ and ξ define the character of twisting around basic line and are defined by (1.1.5);
- function $R(\theta, t)$ defines the character of behavior of radius of main orbit in time and along the basis line and may be arbitrary real smooth functions or some fixed real positive number.
- additional information about known functions and parameters in representation formula (1.3.1) in the static case (i.e. all functions in (1.2.1) do not depend on the time argument) is given in [1].

⋆ In this part we consider displacements which are represented by formula (1.3.1) (or (1.3.1*)) where only one function (parameter) depends on the time argument t and which we shall call **Elementary displacements**.

Correspondingly, according to the previous notations the representation (1.3.1) may be rewritten in following form:

$$X(\tau,\psi,\theta,t) = T_1(t) + [R(\theta,t) + r(\tau,\psi,\theta,t)\cos(\psi + \mu g(\theta) + \xi\omega(t))]\cos(\theta + \varsigma(t))$$
$$Y(\tau,\psi,\theta,t) = T_2(t) + [R(\theta,t) + r(\tau,\psi,\theta,t)\cos(\psi + \mu g(\theta) + \xi\omega(t))]\sin(\theta + \varsigma(t))$$
$$Z(\tau,\psi,\theta,t) = T_3(t) + K(\theta,t) + r(\tau,\psi,\theta,t)\sin(\psi + \mu g(\theta) + +\xi\omega(t) \tag{1.3.1*}$$

where the functions $R(\theta, t)$ defined by (1.1.4*) and $r(\tau, \psi, \theta, t)$ defined by (1.1.2*), depend on time t and on the argument θ;

⋆ A very interesting particular form describes complicate motions of the *GML* bodies with spatial basic line (see Remark 1.3).

$$X(\tau,\psi,\theta,t) = T_1(t) + \cos(\theta + \varsigma(t)) \times \{R(1 + \alpha_1\cos(\omega_1 t)(1 + \beta_1\cos(\kappa_1\theta))+$$
$$+r(1 + \alpha_2\cos(\omega_2 t)(1 + \beta_2\cos(\kappa_2\theta))\cos(\psi + \mu g(\theta) + \xi\omega(t))+$$
$$+\widetilde{r}(1 + \alpha_3\cos(\omega_3 t)(1 + \beta_3\cos(\kappa_3\theta))\cos(\psi + \nu\widetilde{g}(\theta) + \eta\widetilde{\omega}(t))\}$$

$$Y(\tau,\psi,\theta,t) = T_2(t) + \sin(\theta + \varsigma(t)) \times \{R(1 + \alpha_1\cos(\omega_1 t)(1 + \beta_1\cos(\kappa_1\theta))+$$
$$+r(1 + \alpha_2\cos(\omega_2 t)(1 + \beta_2\cos(\kappa_2\theta))\cos(\psi + \mu g(\theta) + \xi\omega(t))+$$
$$+\widetilde{r}(1 + \alpha_3\cos(\omega_3 t)(1 + \beta_3\cos(\kappa_3\theta))\cos(\psi + \nu\widetilde{g}(\theta) + \eta\widetilde{\omega}(t))\}$$

$$Z(\tau,\psi,\theta,t) = T_3(t) + r(1 + \alpha_2\cos(\omega_2 t)(1 + \beta_2\cos(\kappa_2\theta))\sin(\psi + \mu g(\theta) + \xi\omega(t))+$$
$$+ + \widetilde{r}(1 + \alpha_3\cos(\omega_3 t)(1 + \beta_3\cos(\kappa_3\theta))\sin(\psi + \nu\widetilde{g}(\theta) + \eta\widetilde{\omega}(t)). \tag{1.3.1**}$$

General characteristics of Elementary displacements:

We will try to explain, for any geometric change, the role of each function individually in (1.3.1). Assuming that in (1.3.1) only one function depends on the time of the argument, we will describe what movements respectively make this figure. This prompted us to call these movements elementary displacements. If in the previous part of the article we reproduce thc figures and examples of their analytic representation, here we confine ourselves to verbal description. Moving figures in space is spectacular but they can not be transferred simply in a plane. Now we will consider separately the cases, when in (1.3.1) only one function (or vector function) depends on the time of the argument.

Case I. **Vector** $\vec{T}(t)$. Analytic representation (1.3.1) and all its variants (1.3.1*) and (1.3.1**) describe elementary displacements of a geometric object, as a unit, in the space if only vector $(T_1(t), T_2(t), T_3(t))$ depends on the time argument t;

Remark 1.7 Despite a very wide class of movements, we can assume that from a mathematical point of view, these are elementary displacements and during these displacements the geometric shape of initial figures does not change.

Case II. **Function** $K(\theta, t)$. Analytic representation (1.3.1) and all its variants describe elementary displacements of a geometric object that stretches or contracts relative to the axis OZ.

∘ a. If $K(\theta + 2\pi h, t) = K(\theta, t)$, then representations (1.3.1), (1.3.1*) or (1.3.1**) define movements on the Toroids or on the Generalized Möbius-Listing's Body

with spatial basic lines;
o b. If $\dfrac{\partial K(\theta,t)}{\partial t}=0,\ K(\theta,t)=K\cdot\theta\neq 0$, then representations (1.3.1) or (1.3.1*) define a static helix body with constant vertical steps;
o c. If $\dfrac{\partial K(\theta,t)}{\partial t}\neq 0$, then representation (1.3.1) or (1.3.1*) define elementary displacements of some helix like bodies in time along the axis OZ:
(i) **Vertical tension** - if $\dfrac{\partial K(\theta,t)}{\partial t}>0$
(i) **Vertical compression** - if $\dfrac{\partial K(\theta,t)}{\partial t}<0$
o d. If $K(\theta,t)$ is a periodic function of time argument t then representations (1.3.1) or (1.3.1*) define elementary displacements - **Springing figure** with respect to axis OZ: Vertical distances between coils of the figure are changed, or in medical terms body shrinks and expands rhythmically along axis OZ, and function $K(\theta,t)$ defines the **pulsing** or **contraction rhythm** of the organ.

Remark 1.8 **Springing** and **Jumping** both are elementary displacements of figures, but are quite different:

o **Springing** is defined in (Case II.d) and depends on function $K(\theta,t)$ and the vertical distances between coils of figure are changed;

o **Jumping** is a particular case of (Case I). More precisely, if $\dfrac{\partial K(\theta,t)}{\partial t}=0$, $K(\theta,t)\neq 0,\ T_1(t)=T_2(t)\equiv 0$ and only component $T_3(t)\neq 0$ is a periodic function, then representations (1.3.1) or (1.3.1*) define elementary displacements - **jumped helix body** with constant vertical step. Vertical distances between coils of figure are constants, but we have periodic parallel motion with respect to the axis OZ.

Case III. **Functions R_* and R^* depend on time argument**. Elementary displacements which define variations in time of the shape of plane basic line or orbit of the *GTR* figures:

o a. If R_* and R^* in (1.1.4) or R in (1.1.4*) are some smooth functions with respect to time argument t and at the same time 2π-periodic functions of the argument θ, then formula (1.3.1) or (1.3.1*) define a wide set of movements on the toroids or on the helix bodies (face of plane basic line or orbit is a closed plane line, some examples of static variants of this figures are presented in Fig. 8 or 9);
o b. If R_* and R^* in (1.1.4) or R in (1.1.4*) are periodic functions with respect to time argument t, then formulas (1.3.1) or (1.3.1*) define wide sets of movements of the figures with pulsing basic line;

Remark 1.9 If in (1.3.1) only functions R_* and R^* depend on the time argument t, then the shape of the plane basic line of the resulting figures varies in time.

Case IV. **Functions r_* and r^* depend on the time argument**. Elementary displacement which define variations in time of the shapes of radial cross section of the *GTR* figures:

∘ a. If functions r_* and r^* in (1.1.2) depend only on the time argument t, then representation (1.3.1) defines sets of displacements of some point along the lines situated on the surface of the figures, defined by static part of our representation.

∘ b. If r in (1.1.2*) are periodic functions of the time argument t, then representation (1.3.1*) defines set of movements on the bodies with periodical changing radial cross section - in medical term **peristalsis**;

∘ c. If functions r_* and r^* in (1.1.2) (or r in (1.1.2*)) are 2π-periodic functions of ψ, the cross section of the corresponding body is a closed plane figure. In the opposite case, the cross section is open and, in medical terms, one is observing a **rupture** or **scar formation**;

Remark 1.10 ∘ 1. If $D(r)$ - the diameter of radial cross section is smaller than the radius of the space basic line, then representations (1.3.1) or (1.3.1*) define a set of movements on the bodies without self-intersecting points;

∘ 2. This function defines the size, namely maximal and minimal diameters,of the cross section. In practical scenarios, it is not zero (unless we are considering a **sealed** organ). On the other hand, they should be within a reference range (conversely, we might be observing an **aneurysm** or a **burst** in medical terms);

∘ 3. The dependence functions from (1.1.2) (or r in (1.1.2*)) and (1.1.4) (or r in (1.1.4*)) to the variable θ must be in accordance. In particular, if $R(\theta, t)$ is a periodic function of θ, then $r(\tau, \psi, \theta, t)$ has to be periodic with the same period, otherwise the surface of corresponding (1.2.1*) or ((1.3.1) body has a **break**) (example Fig. 10).

Case V. Function $\varsigma(t)$ depends on the time argument. Elementary displacement which defines rotation of the *GTR* figures around of the axis *OZ*;

∘ a. Rule of rotation in the orbit varies in time (or define velocity of twisting), so $\varsigma(t)$ may be an arbitrary smooth function;

∘ b. If $\varsigma(t)$ is some periodic function, then the representation (1.3.1) or all its variants define elementary displacements - Oscillations around of the axis OZ, in medical terms, the **rhythm**.

Case VI. Function $\omega(t)$ depends on the time argument. Elementary displacements which define the twisting of the shape of radial cross section around the basic line of the corresponding *GTR* figures:

∘ a. Rule of twisting around orbit varies in time, so $\omega(t)$ may be arbitrary smooth function;

∘ b. If $\omega(t)$ is some periodic function, then the representation (1.3.1) or all its variants define a wide set of elementary displacements - **Rolling** (periodic twisting) around basic line of figure;

Remark 1.11 This is a very specific movement and it is typical for gases and liquids. In solids such movements are not possible.

Remark 1.12 ⋆ A. If all arguments τ, ψ, θ are fixed constants and the expression depends only on the time argument t, and vector $\overrightarrow{T}(t) = 0$, then representation (1.3.1) or all its variants describe the motion of a particular point along the certain

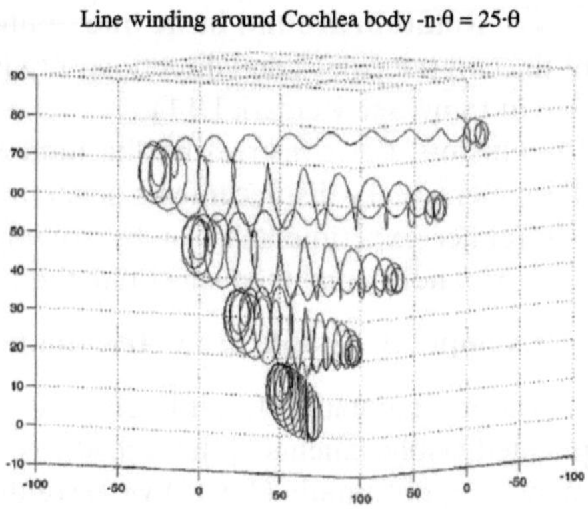

Fig. 12 An example of GTR line

line lying on the surface of the figure, which is defined by same expression. The graph of this function is a space line and this is the well-known **trajectory of displacement of a point**, (for example the lines in Fig. 12);

⋆ B. If τ, θ are fixed constants and the expression depends on the arguments ψ, t and vector $\overrightarrow{T}(t) = 0$, then representation (1.3.1) or all its variants describe the trajectory of displacement of the radial cross section (or some plane figure) (1.1.2) or (1.1.2*) along the basic line of the figure,which is defined by same expression.

Some examples of Complicated displacements:

Example A. Spreading of smoke rings

The opening ceremony of I.G. Petrovski seminar in M. Lomonosov Moscow State University in 1978 started with the lecture of Prof. Mikhail A. Lavrentev. His report about some investigations and problems, which were studied in Novosibirsk, was very interesting. One of these problems is understanding properties of the toroidal waves, which appeared in military techniques and in the atmosphere. He demonstrated one interesting experiment. He put the cigar's smoke in the box with a hole and pushed the opposite side of this box. A **toroidal smoke coil** appeared and it moved forward without degeneration of its structure, but eventually it disappeared. In this chapter we returned to this problem and try to describe model of this process as a superposition of four elementary displacements, i.e. in formula (1.3.1*) we have following terms:

⋆ 1. **Flight**- Vector $(T_1(t), T_2(t), T_3(t)) \neq 0$ defines the character (direction and speed) of the displacement of this coil without degeneration of its structure;

⋆ 2. **Expanding basic line** - Function R in (1.1.4*), according to the model $\dfrac{\partial R(\theta, t)}{\partial t} > 0$ defines the speed of extension of the coil (i.e. of the big radius of torus), but the structure of torus is preserved;

⋆ 3. **Expanding of the radial cross section** - Functions $r(\tau, \psi, \theta)$ in (1.1.2*) define the speed of extension of the shape of radial cross section (i.e. of the small radius of torus);

⋆ 4. **Rotation around basic line** - Function $\omega(t)$ defines the character of twisting of the coil (i.e. speed and direction of twisting of smoke's particles around the basic line of torus, see Remark 1.11).

Conclusion. For constructing the model of this process it is much easier to find these elementary displacements separately, with different views or different methods (either experimentally or by solving the corresponding equations exactly or approximately), and then substitute them into the formula (1.3.1*).

Example B. Simple geometric model of tornado

The representation formula gives us possibility to give a simple model of complicated displacements of the tornado as a combination of four elementary displacements, i.e. in formula (1.3.1*) we have the following terms:

⋆ 1. **Basic movement of tornado** - Vector $(T_1(t), T_2(t), T_3(t)) \neq 0$ and it defines the character (direction and speed) of the displacement of this coil without degeneration of its structure;

⋆ 2. **Rotation** - Function $\varsigma(t)$ defines speed and direction of rotation of the tornado around the vertical OZ axis;

⋆ 3. **Springing** - Function $K(t)$ defines the character of overcoming of big obstacles (mountains and so on ...);

⋆ 4. **Rotation around basic line** - Function $\omega(t)$ defines the character of twisting of the coil (i.e. speed and direction of twisting of of air's particles around the basic helix).

Conclusion. We hope that it will be much easier to measure these elementary components separately and compare against real data using this model.

1.4 Volume of Generalized Twisted and Rotated Bodies

In the general case, the analytical evaluation of the geometrical parameters relevant to complex-shaped organs as functions of time due to arbitrary motions in the three-dimensional space is not a trivial problem. For the specific class of bodies analyzed in this study, the capacity, or volume, may be calculated according to the classic formula:

$$V(t) = \iiint dxdydz = \int_0^{2\pi h} \int_0^{2\pi} \int_{\tau_*}^{\tau^*} \frac{\partial(x, y, z)}{\partial(\tau, \psi, \theta)} d\tau d\psi d\theta, \qquad (1.4.1)$$

involving the Jacobian determinant and without loss of generality, to the simplified calculations we assume that in expression (1.3.1*) numerical parameters $\mu = 1$ and $\xi = 1$:

$$\frac{\partial(x,y,z)}{\partial(\tau,\psi,\theta)} = J(\tau,\psi,\theta) =$$
$$\begin{vmatrix} r_\tau c(\Psi)C(\Theta) & -(R+rc(\Psi))S(\Theta) + (R_\theta + r_\theta c(\Psi) - rg_\theta s(\Psi))C(\Theta) & (r_\psi c(\Psi) - rs(\Psi))C(\Theta) \\ r_\tau c(\Psi)S(\Theta) & (R+rc(\Psi))C(\Theta) + (R_\theta + r_\theta c(\Psi) - rg_\theta s(\Psi))S(\Theta) & (r_\psi c(\Psi) - rs(\Psi))S(\Theta) \\ r_\tau s(\Psi) & K_\theta + r_\theta s(\Psi) + rg_\theta c(\Psi) & r_\psi s(\Psi) + rc(\Psi) \end{vmatrix} \tag{1.4.2}$$

where $r_\tau \equiv \dfrac{\partial r}{\partial \tau}$, $r_\theta \equiv \dfrac{\partial r}{\partial \theta}$, $r_\psi \equiv \dfrac{\partial r}{\partial \psi}$, $c(\Psi) \equiv \cos(\psi + g(\theta) + \omega(t))$, $s(\Psi) \equiv \sin(\psi + g(\theta) + \omega(t))$, $C(\Theta) \equiv \cos(\theta + \varsigma(t))$, and $S(\Theta) \equiv \sin(\theta + \varsigma(t))$. However, after some mathematical manipulations, it is not difficult to derive the expression of the volume of a generalized twisted and rotated body as follows:

$$V(t) = \int_0^{2\pi h}\int_0^{2\pi}\int_{\tau_*}^{\tau^*} (R(\theta,t) + r(\tau,\psi,\theta,t)\cos(\psi + g(\theta) + \omega(t)))$$
$$r(\tau,\psi,\theta,t)\frac{\partial r(\tau,\psi,\theta,t)}{\partial \tau} d\tau d\psi d\theta\,, \tag{1.4.3}$$

or, equivalently:

$$V(t) = \int_0^{2\pi h}\int_0^{2\pi}\int_{\tau_*}^{\tau^*} (R(\theta,t) + r(\tau,\psi,\theta,t)\cos(\psi + g(\theta) + \omega(t)))\, r(\tau,\psi,\theta,t) dr d\psi d\theta\,, \tag{1.4.4}$$

Remark 1.13 If the function $r(\tau,\psi,\theta,t)$ does not depend on argument ψ, the radial cross section of the body under analysis is a circle whose diameter is, in general, function of θ (see Fig. 2 or 3). In this case, since $\int_0^{2\pi} \cos(\psi + g(\theta) + \omega(t))d\psi = 0$, the expression (1.4.3) assumes the classical form [2]:

$$V(t) = \pi \int_0^{2\pi} R(\theta,t)\left[r^2(\tau^*,\theta,t) - r^2(\tau_*,\theta,t)\right] d\theta\,. \tag{1.4.5}$$

1.5 Center of Gravity of Generalized Twisted and Rotated Bodies

Of great interest in practical applications is the center of gravity, whose Cartesian coordinates can be readily evaluated, under the assumption $T_i(t) = 0$ $(i = 1, 2, 3)$, as:

$$x_c(t) = \iiint x dx dy dz,\quad y_c(t) = \iiint y dx dy dz,\quad z_c(t) = \iiint z dx dy dz\,. \tag{1.5.1}$$

Upon combining (1.3.1) and (1.4.4) with (1.5.1), one easily obtains:

$$x_c(t) = \int_0^{2\pi h}\int_0^{2\pi}\int_{\tau_*}^{\tau^*} (R(\theta,t) + r(\tau,\psi,\theta,t)\cos(\psi + g(\theta) + \omega(t)))^2 \cdot \\ \cdot r(\tau,\psi,\theta,t)\cos(\theta + \varsigma(t))dr d\psi d\theta\,, \tag{1.5.2}$$

$$y_c(t) = \int_0^{2\pi h}\int_0^{2\pi}\int_{\tau_*}^{\tau^*} (R(\theta,t) + r(\tau,\psi,\theta,t)\cos(\psi + g(\theta) + \omega(t)))^2 \cdot \\ \cdot r(\tau,\psi,\theta,t)\sin(\theta + \varsigma(t))dr d\psi d\theta\,, \tag{1.5.3}$$

$$z_c(t) = \int_0^{2\pi h}\int_0^{2\pi}\int_{\tau_*}^{\tau^*} (R(\theta,t) + r(\tau,\psi,\theta,t)\cos(\psi + g(\theta) + \omega(t))) \cdot \\ \cdot (K(\theta,t) + r(\tau,\psi,\theta,t)\sin(\psi + g(\theta) + \omega(t)))\, r(\tau,\psi,\theta,t) dr d\psi d\theta\,. \tag{1.5.4}$$

By using Eqs. (1.5.2)–(1.5.4), once can rigorously determine the location of the center of gravity in space and, in this way, infer the type of motion made by the body under analysis.

Remark 1.14 If the functions R, r, g from (1.3.1*) don't depend on argument θ, then the basic line of the body is a circle (with possibly time-dependent diameter), whereas the radial cross section changes only in time with constant twisting rate. In this case, upon noticing that $\int_0^{2\pi} \cos(\theta + \varsigma(t))d\theta = 0$, it is easy to figure out that the coordinates x_c and y_c are identically equal to zero, so that center of gravity can move only along the OZ axis.

Remark 1.15 In case the assumptions in Remarks 1.13 and 1.14 hold true and the function K is either identically equal to zero (the basic line of the body is a circle lying on the OXY plane) or 2π-periodic with respect to the azimuthal angle θ, the coordinate z_c is, also, equal to zero, so that the center of gravity of the body is fixed.

The determination of the surface area of the bodies considered in the presented study is not trivial, since this entails evaluating integral expressions, not reported here for sake of brevity, which involve the first quadratic form of the mapping (1.3.1*) from the τ, ψ, θ coordinates to the Cartesian ones x, y, z, and cannot be reduced to simple formulas.

1.6 Laplace Operator on the Generalized Twisted and Rotated bodies

Many problems of mathematical physics and electromagnetism are related to the Laplace operator. Among them, it is worth mentioning those relevant to the Laplace and Helmholtz, as well as heat and wave equations [13–17] or [3]. In order to address these differential problems in generalized twisted and rotated bodies, a suitable analytical expression of the Laplace operator is helpful. In this contribution, this task is

Table 1 Coefficients of the Laplace operator in different local systems

Terms	Twisted torus	Classical torus	Sphere
$u_{\tau\tau}$	1	1	1
$u_{\theta\theta}$	$(R+\tau\cos(\psi+\mu\theta))^{-2}$	$(R+\tau\cos\psi)^{-2}$	$(\tau\cos\psi)^{-2}$
$u_{\psi\psi}$	$\tau^{-2}+\mu\tau^{-4}(R+\tau\cos(\psi+\mu\theta))^{-2}$	τ^{-2}	τ^{-2}
$u_{\tau\theta}$	0	0	0
$u_{\tau\psi}$	0	0	0
$u_{\theta\psi}$	$-\mu(R+\tau\cos(\psi+\mu\theta))^{-2}$	0	0
u_{τ}	$\frac{1}{\tau}\left(2-\frac{R}{R+\tau\cos(\psi+\mu\theta)}\right)$	$\frac{1}{\tau}\left(2-\frac{R}{R+\tau\cos\psi}\right)$	$\frac{2}{\tau}$
u_{θ}	0	0	0
u_{ψ}	$\frac{-\sin(\psi+\mu\theta)}{\tau(R+\tau\cos(\psi+\mu\theta))}$	$\frac{-\sin\psi}{\tau(R+\tau\cos\psi)}$	$\frac{-\sin\psi}{\tau^2\cos\psi}$

accomplished by introducing two different local coordinate system which directly follow from the representation in (1.3.1*).

First Approach. A system of local coordinates conformal to the "twisted torus" having radius R and twisting parameter μ is introduced, in accordance with (1.3.1*), as follows (it is similar to the spherical coordinates):

$$\begin{cases} x = (R+\tau\cos(\psi+\mu\theta))\cos(\theta))\,, \\ y = (R+\tau\cos(\psi+\mu\theta))\sin(\theta))\,, \\ z = \tau\sin(\psi+\mu\theta)\,, \end{cases} \tag{1.6.1}$$

so that the inverse mapping is readily found to be:

$$\begin{cases} \theta = \arctan\dfrac{y}{x}\,, \\ \tau = \sqrt{z^2+\left(\sqrt{x^2+y^2}-R^2\right)^2}\,, \\ \psi = \arctan\dfrac{z}{\sqrt{x^2+y^2}-R} - \mu\arctan\dfrac{y}{x}\,. \end{cases} \tag{1.6.2}$$

The coefficients of the Laplace operator are listed in the second column of Table 1, in comparison with the corresponding coefficients in classical toroidal, as well as spherical coordinates (Last column of Table 1).

Remark 1.16 1. In the system of local twisted toroidal coordinates, the coefficients of the Laplace equation depend on each of the coordinates τ, ψ, θ. This prevents the application of the eigenfunction method and separation of variables (with respect to τ, ψ, θ) for the solution of differential problems in the considered class of bodies.

2. The coefficients of the Laplace operator strongly depend on the twisting parameter μ. Wherein $\mu = 0$, the mentioned coefficients coincide with those obtained in the classical toroidal coordinate system.

3. Wherein $R = 0$, the coefficients of the Laplace operator coincide with those obtained in the classical spherical coordinate system.

Remark 1.17 The following identities hold true:

$$P(x,y) \equiv x\cos\left(\arctan\frac{y}{x}\right) + y\sin\left(\arctan\frac{y}{x}\right) \equiv \sqrt{x^2+y^2} \equiv R + \tau\cos(\psi + \mu\theta)\,, \tag{1.6.3}$$

$$\begin{gathered}
\frac{\partial P(x,y)}{\partial x} = \frac{x}{x^2+y^2}P(x,y) \equiv \frac{x}{P(x,y)}, \quad \frac{\partial P(x,y)}{\partial y} = \frac{y}{x^2+y^2}P(x,y) \equiv \frac{y}{P(x,y)}, \\
\frac{\partial^2 P(x,y)}{\partial x^2} = \frac{y^2}{(x^2+y^2)^2}P(x,y) \equiv \frac{y^2}{P^3(x,y)}, \quad \frac{\partial^2 P(x,y)}{\partial y^2} = \frac{x^2}{(x^2+y^2)^2}P(x,y) \equiv \frac{x^2}{P^3(x,y)}, \\
\frac{\partial^2 P(x,y)}{\partial x \partial y} = -n\frac{xy}{(x^2+y^2)^2}P(x,y) \equiv -\frac{xy}{P^3(x,y)}.
\end{gathered} \tag{1.6.4}$$

Second Approach. A system of local coordinates conformal to the "twisted torus-like body" having radial cross section with rectangular shape ($0 < \tau < \tau^*$ and $0 < \rho < \rho^*$):

$$\begin{cases}
x = (R + \tau\cos(\mu\theta) - \rho\sin(\mu\theta))\cos(\theta)\,, \\
y = (R + \tau\cos(\mu\theta) - \rho\sin(\mu\theta))\sin(\theta)\,, \\
z = \tau\sin(\mu\theta) + \rho\cos(\mu\theta)\,.
\end{cases} \tag{1.6.5}$$

In this case, the inverse transformation is described by the following equations:

$$\begin{cases}
\theta = \arctan\frac{y}{x}\,, \\
\tau = z\sin\left(\mu\arctan\frac{y}{x}\right) + \left(\sqrt{x^2+y^2} - R\right)\cos\left(\mu\arctan\frac{y}{x}\right)\,, \\
\rho = z\cos\left(\mu\arctan\frac{y}{x}\right) - \left(\sqrt{x^2+y^2} - R\right)\sin\left(\mu\arctan\frac{y}{x}\right)\,,
\end{cases} \tag{1.6.6}$$

so that the expression of the Laplace operator is found to be:

$$\begin{aligned}
\Delta u \equiv{} & \left[1 + \frac{\mu^2\rho^2}{R + \tau\cos(\mu\theta) - \rho\sin(\mu\theta)}\right]\frac{\partial^2 u}{\partial\tau^2} + \left[1 + \frac{\mu^2\tau^2}{R + \tau\cos(\mu\theta) - \rho\sin(\mu\theta)}\right]\frac{\partial^2 u}{\partial\rho^2} + \\
& + \frac{1}{(R + \tau\cos(\mu\theta) - \rho\sin(\mu\theta))^2}\frac{\partial^2 u}{\partial\theta^2} + \frac{\mu\rho}{(R + \tau\cos(\mu\theta) - \rho\sin(\mu\theta))^2}\frac{\partial^2 u}{\partial\tau\partial\theta} + \\
& - \frac{\mu\rho}{(R + \tau\cos(\mu\theta) - \rho\sin(\mu\theta))^2}\frac{\partial^2 u}{\partial\rho\partial\theta} - \frac{\mu\tau\varrho}{(R + \tau\cos(\mu\theta) - \rho\sin(\mu\theta))^2}\frac{\partial^2 u}{\partial\tau\partial\rho} + \\
& + \left[\frac{-\mu^2\tau}{(R + \tau\cos(\mu\theta) - \rho\sin(\mu\theta))^2} + \frac{\cos(\mu\theta)}{R + \tau\cos(\mu\theta) - \rho\sin(\mu\theta)}\right]\frac{\partial u}{\partial\tau} + \\
& + \left[\frac{-\mu^2\rho}{(R + \tau\cos(\mu\theta) - \rho\sin(\mu\theta))^2} - \frac{\sin(\mu\theta)}{R + \tau\cos(\mu\theta) - \rho\sin(\mu\theta)}\right]\frac{\partial u}{\partial\rho}\,.
\end{aligned} \tag{1.6.7}$$

Remark 1.18 1. In the system of local twisted toroidal-like coordinates, the coefficients of the Laplace equation depend on each of the coordinates τ, ρ, θ. This prevents the use of the eigenfunction method and separation of variables (with respect to τ, ρ, θ) for the solution of differential problems in the considered class of bodies.

2. The coefficients of the Laplace operator strongly depend on the twisting parameter μ. Wherein $\mu = 0$ (the radial cross section of the body doesn't twist), the mentioned coefficients coincide with those obtained in the classical cylindrical coordinate system.

1.7 About Some Geometrical Properties of Some Subclasses of Surfaces GTR_2^n

In this section, attention is put on a particular case of "non- regular" generalized twisting and rotated surfaces GTR_2^n described by the following parametric equations:

$$\begin{cases} x(\tau, \psi, \theta) = (R + \tau \cos(\psi + \mu g(\theta))) \cos(\theta), \\ y(\tau, \psi, \theta) = (R + \tau \cos(\psi + \mu g(\theta))) \sin(\theta), \\ z(\tau, \psi, \theta) = K(\theta) + \tau \sin(\psi + \mu g(\theta)), \end{cases} \tag{1.7.1}$$

where R is the radius of orthogonal projection of the basic line of the surface, $g(\theta)$ is defined in (1.1.5) and denotes an arbitrary twisting function with coefficient $\mu = k/2$ ($k \in \mathbb{Z}$), and $K(\theta)$ ia a sufficiently smooth function which affects the vertical stretch of the surface. In (1.7.1) (one particular form of (1.2.1)) the variables τ, ψ, θ are assumed to satisfy the same conditions listed in the Sect. 1.2 for the representation (1.2.1).

After some mathematical manipulations, it can be shown that the tangential vectors of the general surface belonging to the considered class are given by:

$$\vec{\mathbf{r}}_\tau = \begin{Bmatrix} \cos(\psi + \mu g(\theta)) \cos(\theta) \\ \cos(\psi + \mu g(\theta)) \sin(\theta) \\ \sin(\psi + \mu g(\theta)) \end{Bmatrix}, \tag{1.7.2}$$

and:

$$\vec{\mathbf{r}}_\theta = \begin{Bmatrix} -(R + \tau \cos(\psi + \mu g(\theta))) \sin(\theta) - \tau \mu g\prime(\theta) \sin(\psi + \mu g(\theta)) \cos(\theta) \\ (R + \tau \cos(\psi + \mu g(\theta))) \cos(\theta) - \tau \mu g\prime(\theta) \sin(\psi + \mu g(\theta)) \sin(\theta) \\ \tau \mu g\prime(\theta) \cos(\psi + \mu g(\theta)) + K'(\theta) \end{Bmatrix}, \tag{1.7.3}$$

where $g' \equiv \dfrac{\partial g}{\partial \theta}$ and $K' \equiv \dfrac{\partial K}{\partial \theta}$; respectively, so that the expression of the relevant scalar product can be easily derived as:

$$(\vec{\mathbf{r}}_\tau, \vec{\mathbf{r}}_\theta) = K'(\theta) \sin(\psi + \mu g(\theta)). \tag{1.7.4}$$

Remark 1.19 If $K'(\theta) \equiv 0$, then for any integer number k the tangential vectors $\vec{\mathbf{r}}_\tau$, $\vec{\mathbf{r}}_\theta$ are perpendicular to each other, meaning that the local system of coordinates (τ, θ) is orthogonal:

(a) If $K(\theta) \equiv 0$, we have a generalized Möbius-Listing's surface GML_2^n with circular basic line;

(b) If $K(\theta) \equiv const.$, we have a helicoidal surface with constant vertical stretching.

From the expression Eq. (1.7.2), it also follows that:

$$\frac{\partial(x,y)}{\partial(\tau,\theta)} = (R + \tau\cos(\psi + \mu g(\theta)))\cos(\psi + \mu g(\theta)), \tag{1.7.5}$$

$$\begin{aligned}\frac{\partial(z,x)}{\partial(\tau,\theta)} = &-(R + \tau\cos(\psi + \mu g(\theta)))\sin(\psi + \mu g(\theta))\sin(\theta) + \\ &-\left(\tau\mu g'(\theta) - K'(\theta)\cos(\psi + \mu g(\theta))\right)\cos(\theta),\end{aligned} \tag{1.7.6}$$

$$\begin{aligned}\frac{\partial(y,z)}{\partial(\tau,\theta)} = &-(R + \tau\cos(\psi + \mu g(\theta)))\sin(\psi + \mu g(\theta))\cos(\theta) + \\ &+\left(\tau\mu g'(\theta) + K'(\theta)\cos(\psi + \mu g(\theta))\right)\sin(\theta),\end{aligned} \tag{1.7.7}$$

and correspondingly the module of the vector product of these two vectors is found to be:

$$|\vec{\mathbf{r}}_\tau \times \vec{\mathbf{r}}_\theta| = \sqrt{(R + \tau\cos(\psi + \mu g(\theta)))^2 + \left(\tau\mu g'(\theta)/2 + K'(\theta)\cos(\psi + \mu g(\theta))\right)^2}. \tag{1.7.8}$$

Using Eqs. (1.7.5)–(1.7.8), it is not difficult to show that:

Remark 1.20 The following properties of the surfaces (1.7.1) hold true:

(a) If $K(\theta) \neq 0$ and $K(\theta)$ is not a 2π-periodic function, then the corresponding helicoidal surface is two-sided for any integer index k (see (1.7.1) and corresponding $\mu = \frac{k}{2}$);

(b) If $K(\theta) \equiv 0$ or $K(\theta)$ is a 2π-periodic function and k is even number, then the corresponding generalized Möbius-Listing's surface GML_2^k from (1.7.1) is two-sided (because the unit normal vector is a 2π-periodic function);

(c) If $K(\theta) \equiv 0$ or $K(\theta)$ is a 2π-periodic function and k is odd number, then the unit normal vector is a 4π-periodic function and the corresponding generalized Möbius-Listing's surface GML_2^k is one-sided;

The first fundamental form of the general surface belonging to the considered class is described by the equations:

$$E(\tau, \theta) = 1, \tag{1.7.9}$$

$$F(\tau,\theta) = K'(\theta)\sin(\psi + \mu g(\theta)) \ , \tag{1.7.10}$$

$$\begin{aligned} G(\tau,\theta) = {} & (R + \tau\cos(\psi + \mu g(\theta)))^2 + K'^2(\theta)\sin^2(\psi + \mu g(\theta)) + \\ & + \left(\tau\mu g'(\theta) + K'^2(\theta)\cos(\psi + \mu g(\theta))\right)^2 , \end{aligned} \tag{1.7.11}$$

so that:

$$G - F^2 = (R + \tau\cos(\psi + \mu g(\theta)))^2 + \left(\tau\mu g'(\theta) + K'^2(\theta)\cos(\psi + \mu g(\theta))\right)^2 . \tag{1.7.12}$$

Remark 1.21 Each point of a GML_2^k surface (1.7.1) is regular.

The second fundamental form of this class of surfaces is given by:

$$L(\tau,\theta) = 0, \tag{1.7.13}$$

$$M(\tau,\theta) = \frac{2\mu R g'(\theta) - K'(\theta)\cos(\psi + \mu g(\theta))}{\sqrt{EG - F^2}}, \tag{1.7.14}$$

$$\begin{aligned} N(\tau,\theta) = {} & \frac{1}{\sqrt{EG - F^2}}\left\{(R + \tau\cos(\psi + \mu g(\theta)))^2\sin(\psi + \mu g(\theta))\right. \\ & + (R + \tau\cos(\psi + \mu g(\theta)))\left(\tau\mu^2 g''(\theta)) + K''(\theta)\cos(\psi + \mu g(\theta))\right) \\ & \left. + \tau g'(\theta)\left(\tau g'(\theta) + K'(\theta)\cos(\psi + \mu g(\theta))\sin(\psi + \mu g(\theta))\right)\right\} . \end{aligned} \tag{1.7.15}$$

Furthermore, we may rewrite the mean and Gaussian curvatures of this class of surfaces

$$G(\tau,\theta) = \frac{-\left[kRg'(\theta) - K'(\theta)\cos\left(\psi + \frac{kg(\theta)}{2}\right)\right]^2}{\left[EG - F^2\right]^2}. \tag{1.7.16}$$

From expression (1.7.16) it is clear that:

Remark 1.22 The following properties hold true:

(a) Each point of a GTR_2^n surface (1.7.1) is parabolic or hyperbolic (saddle) point;
(b) If $K(\theta) \equiv const.$, then the considered surface features only saddle points.

Acknowledgements This article compiles results from a long standing cooperation and publications of the first author and the co-authors. The results were discussed for a long time and in detail with many colleagues in different countries and institutions. The authors are deeply grateful to Prof. A. Di Nola and D. Gordeziani for a useful discussion of the questions considered in this article. With special gratitude, we want to mark the valuable assistance of Albert Kiefer on graphics. Some parts of the article were written and conceived during visits of Dr. Ilia Tavkhelidze in Rome, in Salerno, in Antwerp and Louvain. W want to thank all the colleagues who participated in debates. Authors are grateful to Ph.D. Student of St. Andrew the First-Called Georgian University of the Patriarchate of Georgia Levan Roinishvili for creating the convenient program for demonstration of the obtained mathematical results in Matlab. Part of the results were obtained with the help of financial support of Shota Rustaveli National Science Foundation (Grant SRNSF/FR/358/5-109/14).

References

1. I. Tavkhelidze, P.E. Ricci, Classification of a wide set of Geometric figures, surfaces and lines (Trajectories), Rendiconti Accademia Nazionale delle Scienze detta dei XL, Memorie di Matematica e Applicazioni, 124°, vol. XXX, fasc. 1, pp. 191–212 (2006)
2. A. Gray, E. Abbena, S. Salamon, *Modern Differential Geometry of Curves and Surfaces with Mathematica*, 3rd edn., Studies in Advanced Mathematics (Chapman & Hall/CRC, Boca Raton, 2006)
3. J. Gielis, D. Caratelli, Y. Fougerolle, P.E. Ricci, I. Tavkhelidze, T. Gerats, Universal natural shapes: from unifying shape description to simple methods for shape analysis and boundary value problems. PLoSOne (2012). doi:10.1371/journal.pone.0029324
4. W. Goemans, I. Van de Woestyne, *Clelia Curves, Twisted Surfaces and Plücker's Conoid in Euclidean and Minkowski 3-Space*, vol. 674, Contemporary Mathematics (2016)
5. V. Targamadze, E. Celesiene, Innovations and its moddeling educationnal enviroment, Bildungswissenschaften und akademisches Selbstverständnis in einer globalisierten, Welt Education and academic self-concept in the globalized world, Gerd-Bodo von Carlsburg, Thomas Vogel Hrsg. (eds.) Frankfurt am Main: Peter Lang. pp. 525–538 (2014)
6. I. Tavkhelidze, Classification of a wide set of trajectories, in *Proceedings of the International Scientific Conference Devoted to 80th Anniversary of Academician I.V. Prangishvili, Information and Computer technologies, Modeling, Control* (Nova Publisher, USA, 2011)
7. I. Tavkhelidze, On the some properties of one class of geometrical figures. TICMI **4**, 51–55 (2000)
8. D. Caratelli, M. Rogava, I. Tavkhelidze, M. Transirico, On some characteristics of the moving generalized Möbius-Listing's bodies. J. Appl. Math. Inf. Mech. **17**(1), 3–15 (2012)
9. J. Gielis, P.E. Ricci, I. Tavkhelidze, About bulky links, generated by generalized Mobius-Listing's bodies GML_n^2. J. Math. Sci. **193**(3), 449–460 (2013)
10. R. Chacon, Modeling natural shapes with a simple nonlinear algorithm. Int. J. Bifurc. Chaos **16**, 2365–2368 (2006)
11. J.R. Munkres, *Elements of Algebraic Topology* (Perseus Books Publisher, New York, 1993)
12. A. Hatcher, *Algebraic Topology* (Cambridge University Press, Cambridge, 2006)
13. D. Caratelli, P.E. Ricci, The Dirichlet problem for the Laplace equation in a starlike domain, in *Proceedings of the International Conference on Scientific Computing*, Las Vegas (2008), pp. 160–166. 14–17 July 2008
14. D. Caratelli, P. Natalini, P.E. Ricci, A. Yarovoy, The Neumann problem for the Helmholtz equation in a starlike planar domain. Appl. Math. Comput. **216**, 556–564 (2010)
15. D. Caratelli, J. Gielis, P.E. Ricci, Fourier-like solution of the Dirichlet problem for the Laplace equation in k-type Gielis domains. J. Pure Appl. Math. Adv. Appl. **5**, 99–111 (2011)
16. D. Caratelli, P. Natalini, P.E. Ricci, J. Gielis, I. Tavkhelidze, The Robin problem for Helmholtz equation in a starlike planar domain. Georgian Math. J. **18**(3), 465–481 (2011)
17. D. Caratelli, P.E. Ricci, J. Gielis, The Robin problem for the Laplace equation in a three-dimensional starlike domain. Appl. Math. Comput. **218**, 713–719 (2011)

Some Properties of "Bulky" Links, Generated by Generalised Möbius–Listing's Bodies $GML_m^n\{0\}$

Ilia Tavkhelidze and Paolo Emilio Ricci

Abstract Natural forms affect all of us, not only for their beauty, but also for their diversity (see e.g. Fig. 1). It is still not known whether forms define the essence of the phenomena associated with them, or vice versa - that is, forms are natural consequences of the phenomena. The essence of one "unexpected" phenomenon is as follows: Usually after one "full cutting", an object is split into two parts. The Möbius strip is a well-known exception, however, which still remains whole after cutting. The first author discovered a class of surfaces, which have following properties - after full cutting more than two surfaces appear, but this is a result for specific class of pure mathematical surfaces [1, 2]. It turns out that three-dimensional Möbius Listing bodies, GML_m^n, which is a wide subclass of the Generalized Twisting and Rotated figures - shortly GTR_m^n - which, through their analytic representation, could yield more than two objects after only single cutting ([3] or [2]). These are not only theoretical results, as can be proved by real-life examples. Many classical objects (torus with different forms of radial cross sections, helicoid, helix, Möbius strip,... etc.) are elements of this wide class of GTR_m^n figures, so it is important to study the similarity and difference between these figures and surfaces. In this chapter we study some questions of similarity and difference in the cases of the "cut" of Generalized Möbius–Listing's figures.

Keywords Analytical representation · Möbius strip · Möbius–Listing's surfaces · Knots · Bulky links

2000 Mathematics Subject Classification 53A05 · 51B10 · 57M25

I. Tavkhelidze (✉)
Department of Mathematics, Iv. Javakhishvili Tbilisi State University, University Street 2, 0186 Tbilisi, Georgia
e-mail: ilia.tavkhelidze@tsu.ge

P.E. Ricci
Department of Mathematics, International Telematic University Uninettuno, Rome, Italy
e-mail: paoloemilioricci@gmail.com

J. Gielis et al. (eds.), *Modeling in Mathematics*,
Atlantis Transactions in Geometry 2, DOI 10.2991/978-94-6239-261-8_11

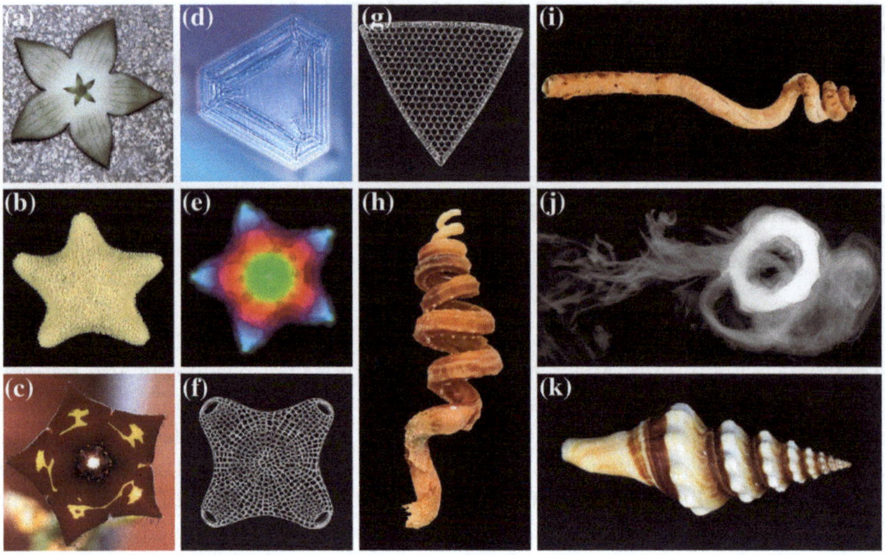

Fig. 1 Examples of Natural and pure geometric figures

1 Introduction

Based on the analytical representation (numeration introduced in the previous chapter of this volume), and on the definition of operation of cutting defined earlier by the first author [1, 2, 4–7], some basic questions to be answered appear, for example:

1. How many objects appear after cutting of the GML_m^n surfaces or bodies?

2. What type of $GML_?^?$ surfaces or bodies appear after cutting (this question for Möbius strip was formulated for the first time by Sosinski see e.g. [8])?

3. What is a link-structure of the surfaces or bodies, which appear after cutting?

4. What are shapes of radial cross sections of the geometric objects which appear after cutting of GML_m^n surfaces or bodies?

5. How many different combinations of geometric objects (in the sense of shapes of the radial cross sections) appear after cutting for arbitrary number m in GML_m^n?

6. What are differential geometric characteristics of GML_m^n surfaces or bodies?

At this stage, we unfortunately do not have answers to all of these questions raised in the case of arbitrary values of m, but some particular cases were reported by the author and his colleagues [1, 2, 5–7].

A tabulation of knots and links of small complexity (thread structure without interior geometry) is well known (see e.g. [8–11] also known as new approaches to the study of these problem, e.g. [12]). In this part of article, we use the analytic representation (1.2.3) (formulas numbering continues from the previous chapter) for the study of bulky knots and links, which appear after a cutting process of the Generalized Möbius–Listing's bodies GML_4^n and GML_5^n along “parallel” lines of their “Ribs” (Def.1.4). In previous articles [1, 2] a set of Generalized Möbius–Listing's surfaces - shortly GML_m^n which are a particular cases of the GTR_m^n bodies (surfaces), have been defined. In particular the case was studied, where the GML_m^n surface (with radial cross section regular a simple star) is cut k-times along lines which are parallel to the basic line. Also, based on the previous results, see e.g. [4–7] we studied the cases when GML_2^n, GML_3^n and GML_4^n are bodies, whose radial cross section are given by a plane 2, 3 or 4-symmetric convex figure with correspondingly 2, 3 or 4 vertices. Now we consider the cases when GML_4^n and GML_5^n, are bodies, whose radial cross section are given by regular square and pentagon correspondingly.

2 General Notations, Definitions and Remarks

In this article, we use the notation, definitions and all numeration introduced in the previous chapter of this volume.

In this part, for simplicity but without loss of generality, we consider Generalized Möbius–Listing's bodies and surfaces GML_m^n with the following restrictions:

• The radial cross sections of these bodies are plane m - symmetric convex figures with m vertices - i.e. $r(\tau, \psi)$ are corresponding Gielis polygons (1.1.3) (for example Fig. 1b, c, e, or g);

• The radial cross sections of these surfaces are plane m-symmetric star like figures with m-vertices - for example:

$$r(\tau, \psi) = \sum_{i=0}^{m-1} l_i \cdot \vartheta\left(\psi - \frac{2\pi}{m} i\right); \tag{2.0.1}$$

where for each $i = 0, \ldots, m-1$, $l_i = const. > 0$ and

$$\vartheta\left(\psi - \frac{2\pi}{m} i\right) \equiv \begin{cases} 0, & if \ \psi \neq \frac{2\pi}{m} i \\ 1, & if \ \psi = \frac{2\pi}{m} i \end{cases}$$

(for more details see in [1] or Fig. 2b, d, or e);

• The basic lines of these $GML_m^n\{\nu\}$ surfaces or bodies, are always torus lines, with characteristic ν defined in (1.1.6) and (1.2.3) (examples are presented in Fig. 1.7). Earlier the cutting along the basic line of $GML_m^n\{0\} \equiv GML_m^n$ always was a circle (see e.g. Fig. 2d or e)!

• The rule of twisting around the basic line is regular. This means in (1.1.5) $g(\theta) \equiv \theta$.

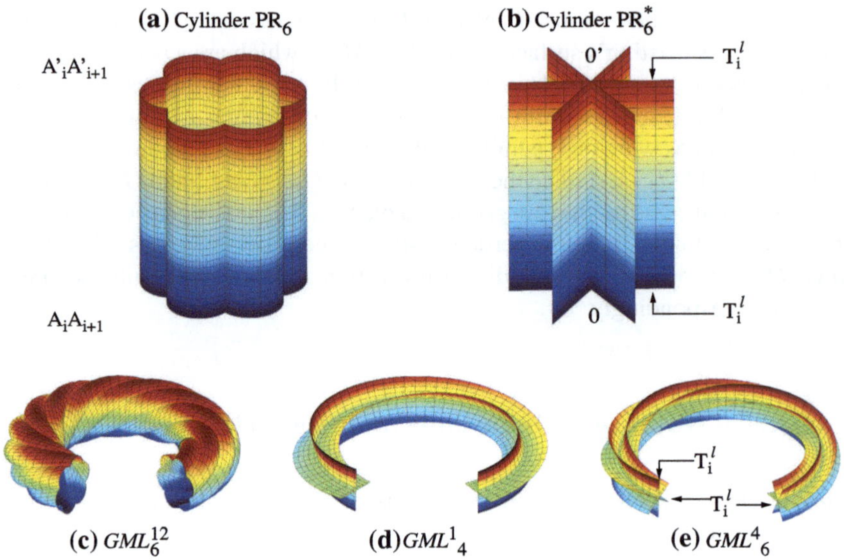

Fig. 2 Examples of the GML surfaces and bodies

According to these restrictions the analytic representations of the corresponding $GML_m^n\{\nu\}$ bodies or surfaces have the form (1.2.3), and we repeat it here:

$$\begin{aligned} X(\tau,\psi,\theta) &= \left[R+\widetilde{R}cos\left(\frac{\widetilde{n}\theta}{\widetilde{m}}\right)+r(\tau,\psi)cos\left(\psi+\frac{n\theta}{m}\right)\right]cos\theta \\ Y(\tau,\psi,\theta) &= \left[R+\widetilde{R}cos\left(\frac{\widetilde{n}\theta}{\widetilde{m}}\right)+r(\tau,\psi)cos\left(\psi+\frac{n\theta}{m}\right)\right]sin\theta \\ Z(\tau,\psi,\theta) &= \widetilde{R}sin\left(\frac{\widetilde{n}\theta}{\widetilde{m}}\right)+r(\tau,\psi)sin\left(\psi+\frac{n\theta}{m}\right). \end{aligned} \tag{2.0.2}$$

Definition 2.1 We call Slit-surface or s-surface of the $GML_m^n\{\nu\}$ body a surface $GML_2^k\{\nu^*\}$ such that:

1. Its basic line is strictly contained within the $GML_m^n\{\nu\}$ body and it is "parallel" to the basic line and ribs of this body;

2. Its radial cross section is a straight line;

3. The line of intersection of the $GML_2^k\{\nu^*\}$ with the $GML_m^n\{\nu\}$ body, which is situated on the side of this body, is "parallel" to the rib line of the $GML_m^n\{\nu\}$ body; This restriction defines the number of rotations k (of surface) which strictly depends on the number of rotation n- of the body;

Definition 2.2 We call Slit-line or s-line of the $GML_m^n\{\nu\}$ surface a space (or plane) closed line (similar to the rib or basic line), which is situated on the $GML_m^n\{\nu\}$ surface and is "parallel" to the basic (or rib's) line (i.e. the distance between this line and the basic or rib's lines is constant).

Definition 2.3 For s-surfaces, without loss of generality, we will use the following notations:

1. $\boldsymbol{S_{1,j}}$**-surface** of the $GML^n_m\{\nu\}$ body is a slit-surface $GML^k_2\{\nu^*\}$ such that the ends of the straight line (radial cross section) are situated on the sides with the numbers 1 (or A_0A_1) and j (or $A_{j-1}Aj$) where $j = 2, 3, \ldots, \left[\frac{m}{2}\right] + 1$; Note that here and always in this article $\left[\frac{p}{q}\right]$ is the integer part of the fraction; corresponding to the plane figures (m symmetric polygon) of the radial cross section of the $GML^n_m\{\nu\}$ body;

2. **SB-surface** of the $GML^n_m\{\nu\}$ body is such $S_{1,[m/2]+1}$ slit-surface, whose radial cross section (straight line) contains the center of symmetry and does not contain vertices of the radial cross section of the $GML^n_m\{\nu\}$ body;

3. $\boldsymbol{VS_{0,j}}$**-surface** of the $GML^n_m\{\nu\}$ body is a slit-surface $GML^k_2\{\nu^*\}$, whose radial cross section (straight line) is situated on the edges with the numbers j (where $j = 2, 3, \ldots, \left[\frac{(m-1)}{2}\right] + 1$) and contains vertex number 0 of the radial cross section of the $GML^n_m\{\nu\}$ body;

4. $\boldsymbol{VBS_{0,[(m-1)/2]+1}}$**-surface** of GML^n_5 body is a slit-surface $GML^k_2\{\nu^*\}$, whose radial cross section (straight line) is situated on the edges with the number $\left[\frac{(m-1)}{2}\right] + 1$, contains the center of symmetry of the polygon and vertex number 0 of the radial cross section of the $GML^n_m\{\nu\}$ body;

5. $\boldsymbol{V_{0,j}}$**-surface** of the $GML^n_m\{\nu\}$ body is a slit-surface $GML^k_2\{\nu^*\}$, whose radial cross sections (straight line) contain correspondingly vertexes numbers 0 and j (where $j = 2, 3, \ldots, \left[\frac{m}{2}\right]$), of the radial cross section of the $GML^n_m\{\nu\}$ body;

Remark 2.1

A. $\boldsymbol{VBS_{0,[(m-1)/2]+1}}$**-surface** - exists only when m (number of symmetry) is a odd number;

B. The radial cross section of the $V_{0,\left[\frac{m}{2}\right]}$- surface contains the center of symmetry of the radial cross section of the $GML^n_m\{\nu\}$ body (polygon) only when m is a even number;

C. According to the regularity of the polygon $P_m \equiv A_0A_1 \ldots A_{m-1}$ it is clear that previous designations are sufficient and do not limit the generality;

D. In the future we will have to consider several cases separately and therefore it is necessary to introduce the following notation:

• 1. $A \equiv |A_0A_1|$ is the full length of the side of the regular polygon (radial cross section of $GML^n_m\{\nu\}$ body);

• 2. $b^1_j \equiv |C^1_j 0 A_1|$ and $b^2_j \equiv |C^2_j 0 A_{j-1}|$ (where $j = 2, 3, \ldots, \left[\frac{(m-1)}{2}\right] + 1$) and points C^1_j and C^2_j are the corresponding ends of straight line $C^1_j C^2_j$ (line is a radial cross section of the corresponding slit-surface $GML^k_2\{\nu^*\}$);

Definition 2.4 A domain, part of the $GML^n_m\{\nu\}$ body (having similar structure to the $GML^k_?\{?\}$ body, usually radial cross sections are not symmetric figures), whose two opposite parallels to the side-surfaces (see Def. 1.4) are slit-surfaces, is called a **Slit zone** or shortly an **s-zone**.

• The thickness of the slit-zone is the distance between two opposite parallel slit-surfaces (i.e. the distance between two opposite parallel straight line in the radial cross section of the slit-zone);

• If the thickness of the slit-zone is zero, then it coincides with a slit-surface.

Without loss of generality, we assume that the thickness of the slit-zone is very small with respect to the size of the body.

Definition 2.5 The "process of cutting" or shortly the "cutting" of a $GML^n_m\{\nu\}$ body is always realized along some s-surface and produces the vanishing (i.e. elimination) of the corresponding s-zone (which possibly reduces to a slit-surface).

• 1. If a $GML^n_m\{\nu\}$ body is cut along an $\boldsymbol{S_{1,j}}$-surface (i.e. $\xrightarrow{S_{1,j}}$), where $j = 2, 3, \ldots, \left[\frac{m}{2}\right] + 1$, then the corresponding vanishing zone is called an $\boldsymbol{S_{1,j}}$**-slit**, and such cutting process is called an $\boldsymbol{S_{1,j}}$ **- zone-slit**;

• 2. If a $GML^n_m\{\nu\}$ body is cut along its $\boldsymbol{SB}$-surface (i.e. $\xrightarrow{SB}$), then the corresponding vanishing zone is called a **SB**-slit, and such cutting process is called an **SB-zone-slit**;

• 3. If $GML^n_m\{\nu\}$ body is cut along its $\boldsymbol{VS_{0,j}}$ -surface (i.e. $\xrightarrow{VS_{0,j}}$), where $j = 2, 3, \ldots, \left[\frac{(m-1)}{2}\right] + 1$ then the corresponding vanishing zone is called a $\boldsymbol{VS_{0,j}}$**-slit**, and such cutting process is called an $\boldsymbol{VS_{0,j}}$**-zone-slit**;

• 4. If a $GML^n_m\{\nu\}$ body is cut along its $\boldsymbol{VBS_{0,[(m-1)/2]+1}}$-surface (i.e. $\xrightarrow{VBS_{0,[(m-1)/2]+1}}$), then the corresponding vanishing zone is called a $\boldsymbol{VBS_{0,[(m-1)/2]+1}}$**-slit**, and such cutting process is called an $\boldsymbol{VBS_{0,[(m-1)/2]+1}}$**-zone-slit**;

• 5. If the $GML^n_m\{\nu\}$ body is cut along its $\boldsymbol{V_{0,j}}$-surface (i.e. $\xrightarrow{V_{0,j}}$), where $j = 2, 3, \ldots, \left[\frac{m}{2}\right]$, then the corresponding vanishing zone is called a $\boldsymbol{V_{0,j}}$ – **slit**, and such cutting process is called an $\boldsymbol{V_{0,j}}$**-zone-slit**;

Remark 2.2

∘ A. For completeness, for the review of all cases of $\boldsymbol{S_{1,j}}$-zone-slit, three different variants should be considered separately for each value of j=2, 3, ..., $\left[\frac{(m-1)}{2}\right]+1$:

$$1.\ b^1_j + b^2_j < A; \qquad 2.\ b^1_j + b^2_j = A; \qquad 3.\ \ b^1_j + b^2_j > A;$$

∘ B. In particular, for $\boldsymbol{S_{1,[m/2]+1}}$-zone-slit, when m is odd number, five different variants should be considered separately:

⋆ 1. $b^1_{[m/2]+1} + b^2_{[m/2]+1} < A$ and one of the number $b^i_{[m/2]+1} \le 0.5A,\ \ i = 1$ or 2;

⋆ 2. $b^1_{[m/2]+1} + b^2_{[m/2]+1} = A$;

★ 3. $b^1_{[m/2]+1} + b^2_{[m/2]+1} > A$ and one of the number $b^i_{[m/2]+1} \leq 0.5A,\ i = 1$ or 2;

★ 4. $b^1_{[m/2]+1} + b^2_{[m/2]+1} < A$ and both of the numbers $b^i_{[m/2]+1} > 0.5A,\ i = 1$ or 2;

★ 5. We consider separately the case when the line $C^1_{[m/2]+1}C^2_{[m/2]+1} < A$ contains a center of symmetry of the polygon and correspondingly we have a SB-zone-slit.

2.1 Relations Between the Set of Generalized Möbius–Listing's Surfaces $GML^n_m\{\nu\}$ and the Sets of Ribbon Knots and Ribbon Links

In this part we recall the results obtained for the $GML^n_m\{\nu\}$ surfaces. It is necessary for comparisons of results for $GML^n_m\{\nu\}$ bodies and surfaces, and for the proof of the corresponding theorems. Without loss of generality, we will consider Generalized Möbius–Listing's Surfaces with radial cross section a simple star (2.0.1), perhaps with different lengths of the wings. We recall definitions (Definitions 2.6–2.9) introduced for surfaces in earlier articles [1, 2]. It should be noted that the results were obtained in these articles for $GML^n_m\{\nu\}$ surfaces for any value of the numbers m and n, and the number of cuts k. But now we present the formulation of the theorem for a single cut.

Definition 2.6 (*analogous to slit-surface Definition* 2.1) A closed line (similar to the basic or border's line) which is situated on a $GML^n_m\{\nu\}$ surfaces and is "parallel" to the basic (or border's) line of the $GML^n_m\{\nu\}$, i.e. the distances between this line and basic or border's lines are constants, is called a **Slit line** or shortly "s-line".

∘ If the distance between an s-line and the basic line is zero, then this s-line coincides with the basic line (and sometimes is called "B-line").

Definition 2.7 A domain situated on the surface $GML^n_m\{\nu\}$ and such that its border lines are slit lines, is called a "Slit-zone" or shortly an "s-zone".

The distance between the border's lines of an s-zone is the "width" of this s-zone.

∘ If the width of the s-zone is equal to zero, then this zone reduces to an s-line.

Definition 2.8 If the "B-line" is properly contained inside a "Slit-zone" - i.e. the distance to the border's lines is strictly positive, then this "Slit-zone" will be called a "B-zone".

Definition 2.9 The "process of cutting" of the $GML^n_m\{\nu\}$ surfaces or shortly the "cutting" is always realized along some s-lines and produces the vanishing (i.e. elimination) of the corresponding s-zone (which in case reduces to an s-line).

∘ If a $GML^n_m\{\nu\}$ surface is cut along an s-line, then the resulting object is called "slitting" and the corresponding vanishing zone will be called an s-slit.

∘ If a $GML^n_m\{\nu\}$ surface is cut along its B-line, then the resulting object is called "B-slitting" and the corresponding vanishing zone will be called a B-slit.

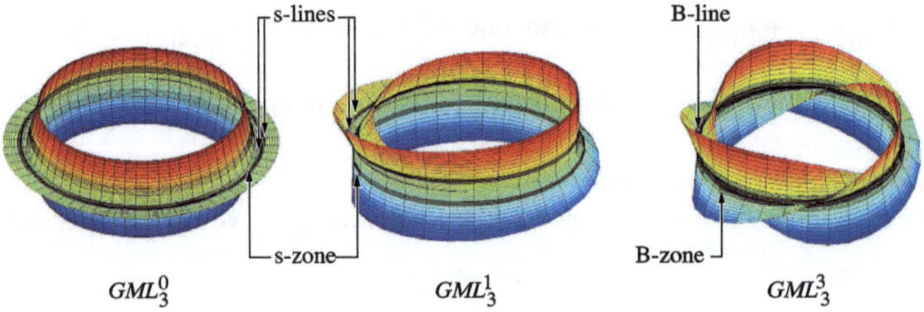

Fig. 3 Examples of s-lines

∘ If the vanishing zone after an s-slit (a B-slit) is given by an "s-zone" (a "B-zone"), then the cutting process will be called an **s-zone-slit** (a **B-zone-slit**) see examples in Fig. 3.

Sometimes in this article we use "link-1" to the notation of "Knot". However, "Links" are always correspondingly "Ribbon Links" if we consider surfaces and "Bulky Links" if we consider $GML_m^n\{\nu\}$ bodies!

Theorem 1 *If the number of twisting is $n \equiv m\omega$, where $\omega \in \mathbb{Z}$ (ω is the number of full rotations of the radial cross section of a body around its basic line) and the $GML_m^n\{0\} \equiv GML_m^n$ surface is cut along some of its slit-lines, then:*

⋆ ***Case A.*** *- after an S-zone-slit (i.e. a line which does not coincide with its basic line) of the GML_m^n surface, always appears an object **ribbon bulk link-2** $\{(2\omega)_1^2\}$ (according to the classic tabulation of the links of small complexity [9, 10]) of the $GML_m^{m\omega}\{0\}$ and $GML_m^n\{\omega\}$ surfaces, whose radial cross section are correspondingly simple star with m wings (with different lengths) and line (Möbius like surface), i.e. for each natural $\omega = 0, 1, 2, \ldots$*

$$GML_m^{m\omega} \xrightarrow{S} link - 2\{(2\omega)_1^2\}\ of\ the\ GML_2^{2\omega}\{\omega\}\ and\ GML_m^{m\omega}\{0\} \qquad (2.1.1)$$

⋆ ***Case B.*** *- after a B-zone-slit (i.e. s-line coincides with its basic line) of the GML_m^n surface, appears an object **ribbon bulk link-**m of the m different geometric $GML_2^{2\omega}\{\omega\}$ surfaces, whose radial cross section are correspondingly lines (Möbius like surface), i.e. for each natural $\omega = 0, 1, 2, \ldots$*

$$GML_m^{m\omega} \xrightarrow{B} link - m\ of\ the\ m\ surfaces\ GML_2^{2\omega}\{\omega\} \qquad (2.1.2)$$

Theorem 2 *If the number of twisting is $n = m \cdot \omega + q$, $q = 1, 2, \ldots, m-1$, where $\omega \in \mathbb{Z}$; $\varsigma = \gcd(m, q)$ (i.e. $m = \varsigma \cdot \xi$ and $q = \varsigma \cdot \eta$ and $\xi, \eta \in \mathbb{Z}$) and the GML_m^n surface is cut along some of its slit-lines, then:*

⋆ ***Case A.*** *- after an S-zone-slit (i.e. a line which does not coincide with its basic line) of the GML_m^n appears an object **ribbon bulk link-2** of the $GML_m^n\{0\}$ and $GML_2^{2\cdot\xi(\omega+1)+2(\eta-1)}\{\omega + \dfrac{q}{m}\}$ surfaces. The first one has the same structure $(0)_1$ as* *the initial surface before cutting; whose radial cross section are correspondingly simple star with m wings (with different lengths) and line (Möbius like surface) i.e. for each natural* $\omega = 0, 1, 2, \ldots$

$$GML_m^{m\omega+q} \xrightarrow{S} link - 2\ of\ the\ GML_m^{m\omega}\{0\}\ and\ GML_2^{2\cdot\xi(\omega+1)+2(\eta-1)}\{\omega + \frac{q}{m}\} \quad (2.1.3)$$

⋆ ***Case B.*** *- after a B-zone-slit (i.e. s-line coincides with its basic line)of the* GML_m^n *surface, appears an object* ***ribbon bulk link-***ς *of the* ς *different geometric* $GML_2^{2\cdot\xi(\omega+1)+2(\eta-1)}\{\omega + \frac{q}{m}\}$ *surfaces, whose radial cross sections are correspondingly lines (Möbius like surface), i.e. for each natural* $\omega = 0, 1, 2, \ldots$

$$GML_m^{m\omega+q} \xrightarrow{B} link - \varsigma\ of\ the\ \varsigma\ surfaces\ GML_2^{2\cdot\xi(\omega+1)+2(\eta-1)}\{\omega + \frac{q}{m}\} \quad (2.1.4)$$

2.2 Relations Between the Set of Generalized Möbius–Listing's Bodies $GML_4^n\{\nu\}$ and the Sets of Bulky Knots and Bulky Links

Without loss of generality and to simplify the proofs of theorems, we will consider initial (before cutting) Generalized Möbius–Listing's Body $GML_4^n\{0\} \equiv GML_4^n$, whose radial cross section is a square (corresponding parameters in Gielis Superformula for (1.1.3) function $r(\tau, \theta)$) and its basic line (see Definition 1.3 from the previous chapter) is a circle. This means that the analytic representation (2.0.2) of corresponding body is simpler and has following form

$$\begin{cases} X(\tau,\theta) = \left(R + r(\tau,\theta)\cos\left(\psi + \frac{n\theta}{4}\right)\right)\cos(\theta)), \\ Y(\tau,\theta) = \left(R + r(\tau,\theta)\cos\left(\psi + \frac{n\theta}{4}\right)\right)\sin(\theta)), \\ Z(\tau,\theta) = r(\tau,\theta)\sin\left(\psi + \frac{n\theta}{4}\right), \end{cases} \quad (2.2.1)$$

Using the technique described in [1, 5, 6] we obtain the following theorems:

Theorem 3 *If the number of twisting is* $n = 4\omega$, *where* $\omega \in \mathbb{Z}$ *(*ω *is the number of full rotations of radial cross section of a body around its basic line) and the* GML_4^n *body is cut along some of its slit-surfaces, then always an object "bulk link-2"* $\{(2\omega)_1^2\}$ *(according to the classic tabulation of the links of small complexity [9–11]) of the two bulk link-1 appears; But both components of this bulk link-2 have four different geometric structures, more precisely:*

⋆ ***Case A.*** *- after an* $S_{1,2}$*-zone-slit of the* GML_4^n *body, an object bulk link-2* $\{(2\omega)_1^2\}$ *of the* $GML_3^{3\omega}\{\omega\}$ *and* $GML_5^{5\omega}\{0\}$ *bodies appears, whose radial cross sections are correspondingly 3 and 5 angular plane figures, i.e. for each natural* $\omega = 0, 1, 2, \ldots$

$$GML_4^{4\omega} \xrightarrow{S_{1,2}} link-2\{(2\omega)_1^2\}\ of\ the\ GML_3^{3\omega}\{\omega\}\ and\ GML_5^{5\omega}\{0\} \tag{2.2.2}$$

⋆ ***Case B.*** *- after an $S_{1,3}$-zone-slit of the GML_4^n body, an object bulk link-2 $\{(2\omega)_1^2\}$ of the $GML_4^{4\omega}\{\mu\}$ and $GML_4^{4\omega}\{\nu\}$ bodies appears, whose radial cross section are four angular, but usually not identical plane figures, i.e. for each natural $\omega = 0, 1, 2, \ldots$*

$$GML_4^{4\omega} \xrightarrow{S_{1,3}} link-2\{(2\omega)_1^2\}\ of\ the\ GML_4^{4\omega}\{\mu\}\ and\ GML_4^{4\omega}\{\nu\} \tag{2.2.3}$$

There are two fundamentally different subcases, that are associated with finding the center of symmetry of the initial figures after the cut; more precisely - if the center of symmetry of the radial cross section (initial polygon) after cutting remains in:

⋆ *a. one the four-angular parts of the domain, then that figure, after the cut will have a characteristic $\nu = 0$, i.e. in formula (2.2.3) $\mu = \omega$ and $\nu = 0$;*

⋆ *b. the cutting line (i.e. SB-slit), then in formula (2.2.3) both element have characteristic $\mu = \nu = \omega$. In this subcase both geometric objects are identic.*

Case C. *- after an $VS_{0,2}$-zone-slit of the GML_4^n body an object bulk link-2 $\{(2\omega)_1^2\}$ of the $GML_3^{3\omega}\{\omega\}$ and $GML_4^{4\omega}\{0\}$ bodies appears, whose radial cross section are correspondingly 3 and 4 angular plane figures, i.e. for each natural $\omega = 0, 1, 2, \ldots$*

$$GML_4^{4\omega} \xrightarrow{VS_{0,2}} link-2\{(2\omega)_1^2\}\ of\ the\ GML_3^{3\omega}\{\omega\}\ and\ GML_4^{4\omega}\{0\} \tag{2.2.4}$$

Case D. *- after an $V_{0,2}$-zone-slit of the GML_4^n body an object bulk link-2 $\{(2\omega)_1^2\}$, whose both components are $GML_3^{3\omega}\{\omega\}$ identic bodies, appears, i.e. for each natural $\omega = 0, 1, 2, \ldots$*

$$GML_4^{4\omega} \xrightarrow{V_{0,2}} link-2\{(2\omega)_1^2\}\ of\ the\ two\ GML_3^{3\omega}\{\omega\} \tag{2.2.5}$$

Sketch of the proof. We draw straight lines connecting the center of symmetry with the corresponding vertices, so we obtain a star like cross, and this case of GML_4^n, generalized Möbius–Listing's surfaces, has already been studied (see Theorem 1 and full text of proof in [1] or [2]). Slit surface for bulky links is converted in a corresponding slit-line, but in this situation we know all possible variants that appear after cutting. As can be seen from Theorem 1, there are two fundamentally different cases:

⋆ 1. When the s-line does not contain the center of symmetry of the radial cross section of the initial figure (before cutting, Theorem 1 case A (2.1.1)) - corresponding cases for GML_4^n body are (cases A., B.a. and C. in Theorem 3);

⋆ 2. In the opposite case, B-line contains this point (Theorem 1. case B (2.1.2)) corresponding cases for GML_4^n body are (cases B.b. and D.).

Corollary 1 *In the case of GML bodies in addition there are actual distinctions of shapes of plane elements of the radial cross section of figures which appear after cutting. The form of "new figures" (i.e. figures which appear after the cutting*

Shapes of radial cross sections $GML_4^{4\omega}$	-slit	Parameters of the objects which appear after cutting			
		Shapes of radial cross sections	Structure of elements	Link group of elements	Link group of object
A. 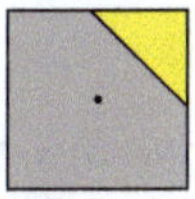	$S_{1,2}$	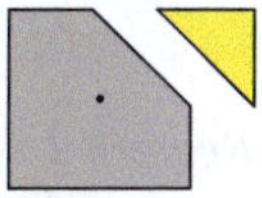	$GML_5^{5\omega}$ $GML_3^{3\omega}\{\omega\}$		
B. a. 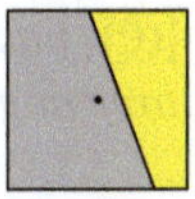	$S_{1,3}$	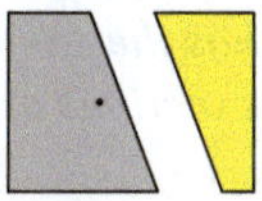	$GML_4^{4\omega}$ $GML_4^{4\omega}\{\omega\}$	Link-1	Link-2
B. b. 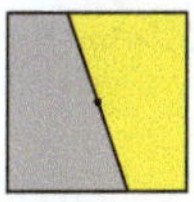	SB	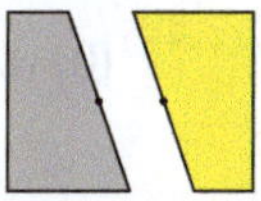	$GML_4^{4\omega}\{\omega\}$ $GML_4^{4\omega}\{\omega\}$		
C. 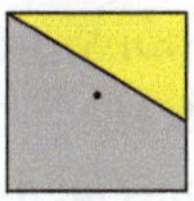	$SV_{0,2}$	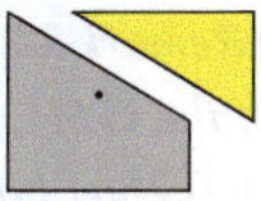	$GML_4^{4\omega}$ $GML_3^{3\omega}\{\omega\}$	$\{0_1\}$	$\{(2\omega)_1^2\}$
D. 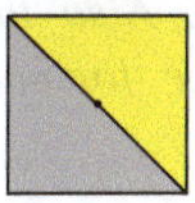	$V_{0,2}$	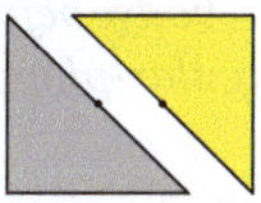	$GML_3^{3\omega}\{\omega\}$ $GML_3^{3\omega}\{\omega\}$		

Fig. 4 Table of all possible cases and corresponding shapes of the radial cross sections which appear after the cutting process of GML bodies, with symmetry number $m = 4$ and number of rotations multiples of 4

process) depends on the location of the cut surface (more precisely the location of the corresponding line in the radial cross section)!

Finally we return to the case under consideration and we count separately how many rotations make some vertices of new bodies that appear after cutting!

All possible cases and corresponding shapes of the radial cross sections after the cutting process are given in the table in Fig. 4.

Theorem 4 *If the number of twisting is $n = 4 \cdot \omega + q$, $q = 1$ or 3, where $\omega \in \mathbb{Z}$ and the GML_4^n body is cut along some of its slit-surfaces, then **seven different cases** with different geometric forms of the radial cross sections appear, more precisely the following results hold:*

⋆ ***Case A.*** *Taking into account Remark 2.2.A, we have three different subcases:*

∘ ***Subcase A.I.*** *if $b_2^1 + b_2^2 < A$; (see Remark 2.2-A.1), after an $S_{1,2}$-zone-slit of the GML_4^n body an object bulk link-2, of the $GML_8^{2(4\omega+q)}\{0\}$ and $GML_3^{3(4(\omega+1)+(q-1))}$ $\{\omega + 0.25q\}$ bodies appear. The first one has the same structure $\{(0)_1\}$ as the initial body before cutting. Also their radial cross sections are eight and three angular plane* *figures (usually there are not regular polygons), i.e. for each natural* $\omega = 0, 1, 2, \ldots$

$$GML_4^{4\omega+q} \xrightarrow{S_{1,2}} link-2\ of\ the\ GML_8^{2(4\omega+q)}\{0\}\ and\ GML_3^{12\omega+3q+9}\left\{\omega+\frac{q}{4}\right\} \tag{2.2.6}$$

∘ ***Subcase A.II.*** *if* $b_2^1 + b_2^2 = A$*; (see Remark 2.2-A.2), after an* $S_{1,2}$*-zone-slit of the* GML_4^n *body an object bulk link-2, of the* $GML_4^{(4\omega+q)}\{0\}$ *and* $GML_3^{3(4(\omega+1)+(q-1))}$ $\{\omega+0.25q\}$ *bodies appear. The first one has the same structure* $\{(0)_1\}$ *as the initial body before cutting. Also their radial cross sections are four and three angular plane figures, i.e. for each natural* $\omega = 0, 1, 2, \ldots$

$$GML_4^{4\omega+q} \xrightarrow{S_{1,2}} link-2\ of\ the\ GML_4^{(4\omega+q)}\{0\}\ and\ GML_3^{12\omega+3q+9}\left\{\omega+\frac{q}{4}\right\} \tag{2.2.7}$$

∘ ***Subcase A.III.*** *if* $b_2^1 + b_2^2 > A$*; (see Remark 2.2-A.3), after an* $S_{1,2}$*-zone-slit of the* GML_4^n *body an object bulk link- 3, of the* $GML_4^{(4\omega+q)}\{0\}$*,* $GML_3^{3(4(\omega+1)+(q-1))}$ $\{\omega+0.25q\}$ *and* $GML_5^{5(4(\omega+1)+(q-1))}\{\omega+0.25q\}$ *bodies appear. The first one has the same structure* $\{(0)_1\}$ *as the initial body before cutting. Also their radial cross sections are four, three and five angular plane figures, i.e. for each natural* $\omega = 0, 1, 2, \ldots$

$$\begin{aligned} &GML_4^{4\omega+q} \xrightarrow{S_{1,2}} link-3\ of\ the\ GML_4^{(4\omega+q)}\{0\}\ , \\ &GML_3^{12\omega+3q+9}\left\{\omega+\frac{q}{4}\right\}\ and\ GML_5^{20\omega+5q+15}\left\{\omega+\frac{q}{4}\right\} \end{aligned} \tag{2.2.8}$$

⋆ ***Case B.*** *Taking into account Remark 2.2.B, we have two different subcases:*

∘ ***Subcase B.I.*** *after an* $S_{1,3}$*-zone-slit (which is not SB-zone-slit (see Remark 2.2.B.5 and Definition 2.5) of the* GML_4^n *body an object bulk link-3, of the* $GML_4^{(4\omega+q)}\{0\}$ *and two* $GML_4^{4(4(\omega+1)+(q-1))}\{\omega+0.25q\}$ *bodies appear. The first one has the same structure* $\{(0)_1\}$ *as the initial body before cutting. Also their radial cross sections are four (not identic) angular plane figures, i.e. for each natural* $\omega = 0, 1, 2, \ldots$

$$GML_4^{4\omega+q} \xrightarrow{S_{1,3}} link-3\ of\ the\ GML_4^{(4\omega+q)}\{0\}\ ,\ and\ two\ GML_4^{16\omega+4q+12}\left\{\omega+\frac{q}{4}\right\} \tag{2.2.9}$$

∘ ***Subcase B.II.*** *after an* SB*-zone-slit of the* GML_4^n *body an object bulk link-1 (or knot), of the* $GML_4^{4(4(\omega+1)+(q-1))}\{\omega+0.25q\}$ *body appears. Its radial cross section is given by four angular plane figures, i.e. for each natural* $\omega = 0, 1, 2, \ldots$

$$GML_4^{4\omega+q} \xrightarrow{SB} link-1\ of\ the\ GML_4^{16\omega+4q+12}\left\{\omega+\frac{q}{4}\right\} \tag{2.2.10}$$

⋆ ***Case C.*** *after an $VS_{0,2}$-zone-slit of the GML_4^n body (see Definition 2.5) an object bulk link-3, of the $GML_4^{(4\omega+q)}\{0\}$, $GML_3^{3(4(\omega+1)+(q-1))}\{\omega+0.25q\}$ and $GML_4^{4(4(\omega+1)+(q-1))}\{\omega+0.25q\}$ bodies appear, whose radial cross sections are correspondingly four, three and four angular plane figures, i.e. for each natural* $\omega = 0, 1, 2, \ldots$

$$\begin{aligned} &GML_4^{4\omega+q} \overset{VS_{0,2}}{\longrightarrow} link-3\ of\ the\ GML_4^{(4\omega+q)}\{0\}\,, \\ &GML_3^{12\omega+3q+9}\left\{\omega+\frac{q}{4}\right\}\ and\ GML_4^{16\omega+4q+12}\left\{\omega+\frac{q}{4}\right\} \end{aligned} \tag{2.2.11}$$

⋆ ***Case D.*** *after an $V_{0,2}$-zone-slit of the GML_4^n body an object bulk link-1 (or knot), of the $GML_3^{3(4(\omega+1)+(q-1))}\{\omega+0.25q\}$ body appears. Its radial cross section is given by three angular plane figure, i.e. for each natural $\omega = 0, 1, 2, \ldots$*

$$GML_4^{4\omega+q} \overset{V_{0,2}}{\longrightarrow} link-1\ of\ the\ GML_3^{12\omega+3q+9}\left\{\omega+\frac{q}{4}\right\} \tag{2.2.12}$$

Sketch of the proof. For the proof the previous argument can be repeated verbatim (see Theorem 3), but it is necessary to use the corresponding results of Theorem 2, when the greatest common divisor of $m = 4$ and q is equal to 1!

As can be seen from Theorem 2, there are two fundamentally different cases:

⋆ 1. When the s-line does not contain the center of symmetry of the radial cross section of the initial figure (before cutting, Theorem 2. case A (2.1.3)) - corresponding cases for GML_m^n body are (cases A.I., A.II., A.III., B.I., and C.);

⋆ 2. In the opposite case, the b-line contains this point (Theorem 2. case B (2.1.4)) - corresponding cases for body are (cases B.II. and D.).

Corollary 2 *In the case of GML bodies in addition there are actual distinctions of shapes of plane elements of the radial cross section of figures which appear after cutting. The form of "new figures" (i.e. figures which appear after the cutting process) depends on the location of the cut surface (more precisely the location of the corresponding line in the radial cross section)!*

⋆ ***A new effect is that the s-surface can be self-intersecting, and therefore appear in more than two objects*** *(see cases A.III., B.I. and C.)!*

⋆ *But also the Möbius phenomenon is preserved - i.e. after a full cut only one object appears (see cases B.II., and D.)!*

Finally, we return to the case under consideration and we count separately numbers of rotations of vertices around basic lines of the new bodies that appear after cutting!

All possible cases and corresponding shapes of the radial cross sections after the cutting process are given in the table in Fig. 5.

Theorem 5 *If the number of twisting is $n = 4 \cdot \omega + 2$, where $\omega \in \mathbb{Z}$ and the GML_4^n body is cut along some of its slit-surfaces, then* ***five different cases*** *with different*

Shapes of radial cross section GML_4^n	-slit	Parameters of the objects which appear after cutting		
		Shapes of radial cross sections	Structure of elements if $n = 4\omega + q$	Link group of object
A.I 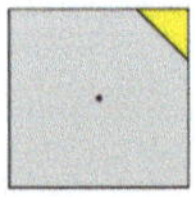	$S_{1,2}$ $b_2^1+b_2^2<A$	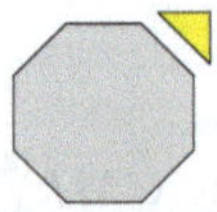	$GML_8^{8\omega+2q}$ $GML_3^{12\omega+3q+9}\{\omega+0.25q\}$ 	Link-2
A.II 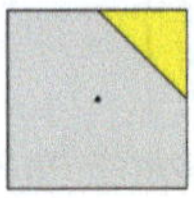	$S_{1,2}$ $b_2^1+b_2^2=A$	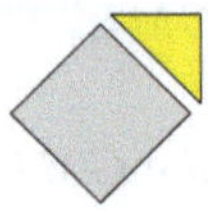	$GML_4^{4\omega+q}$ $GML_3^{12\omega+3q+9}\{\omega+0.25q\}$ 	Link-2
A.III 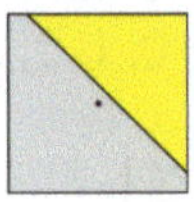	$S_{1,2}$ $b_2^1+b_2^2>A$		$GML_4^{4\omega+q}$ $GML_5^{20\omega+5q+15}\{\omega+0.25q\}$ $GML_3^{12\omega+3q+9}\{\omega+0.25q\}$ 	Link-3
B.I 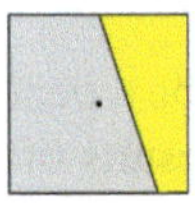	$S_{1,3}$		$GML_4^{4\omega+q}$ $GML_4^{16\omega+4q+12}\{\omega+0.25q\}$ $GML_4^{16\omega+4q+12}\{\omega+0.25q\}$ 	Link-3
B.II 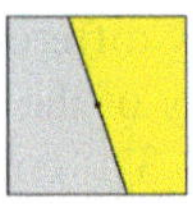	SB		$GML_4^{16\omega+4q+12}\{\omega+0.25q\}$ 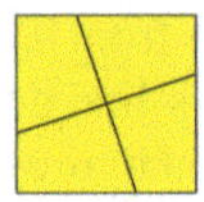	Link-1
C. 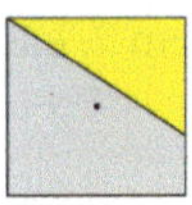	$VS_{0,2}$	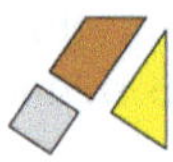	$GML_4^{4\omega+q}$ $GML_4^{16\omega+4q+12}\{\omega+0.25q\}$ $GML_3^{12\omega+3q+9}\{\omega+0.25q\}$ 	Link-3
D. 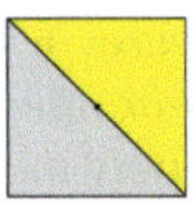	$V_{0,2}$		$GML_3^{12\omega+3q+9}\{\omega+0.25q\}$ 	Link-1

Fig. 5 Table of all possible cases and corresponding shapes of the radial cross sections which appear after the cutting process of GML bodies, with symmetry number $m = 4$ and odd number of rotations

geometric forms of the radial cross sections appear. More precisely following results hold:

⋆ ***Case A.*** *after an $S_{1,2}$-zone-slit of the GML_m^n body, an object bulk link-2, of the $GML_6^{3(2\omega+1)}\{0\}$ and $GML_3^{3\cdot 2(\omega+1)}\{\omega+0.5q\}$ bodies appear. The first one has the same structure $\{(0)_1\}$ as the initial body before cutting. Also their radial cross sections are six and three angular plane figures (usually there are not regular polygons), i.e. for each natural $\omega = 0, 1, 2, \ldots$*

$$GML_4^{4\omega+2} \xrightarrow{S_{1,2}} link-2\ of\ the\ GML_6^{(6\omega+3)}\{0\}\ and\ GML_3^{6\omega+6}\left\{\omega+\frac{1}{2}\right\} \tag{2.2.13}$$

⋆ ***Case B.*** *Taking into account Definition 2.5. and Remark 2.2-B.5, we have two different subcases:*

∘ ***Subcase B.I.*** *after an $S_{1,3}$-zone-slit (which is not a SB-zone-slit) of the GML_4^n body, an object bulk link-2, of the $GML_4^{2(2\omega+1)}\{0\}$ and $GML_4^{4\cdot2(\omega+1)}\{\omega+0.5q\}$ bodies appear. The first one has the same structure $\{(0)_1\}$ as the initial body before cutting. Also their radial cross sections are four angular (not identic) plane figures, i.e. for each natural $\omega = 0, 1, 2, \ldots$*

$$GML_4^{4\omega+2} \xrightarrow{S_{1,3}} link-2\ of\ the\ GML_4^{(4\omega+2)}\{0\}\ and\ GML_4^{8\omega+8}\left\{\omega+\frac{1}{2}\right\} \tag{2.2.14}$$

∘ ***Subcase B.II.*** *after an SB-zone-slit of the GML_4^n body, an object bulk link-1 (knot), of the $GML_4^{4\cdot2(\omega+1)}\{\omega+0.5q\}$ body appears. Its radial cross section is given by four angular plane figures, i.e. for each natural $\omega = 0, 1, 2, \ldots$*

$$GML_4^{4\omega+2} \xrightarrow{BS} link-1\ of\ the\ GML_4^{8\omega+8}\left\{\omega+\frac{1}{2}\right\} \tag{2.2.15}$$

⋆ ***Case C.*** *after an $VS_{0,2}$-zone-slit of the GML_4^n body (see Definition 2.5), an object bulk link-2, of the $GML_4^{2(2\omega+1)}\{0\}$ and $GML_3^{3\cdot2(\omega+1)}\{\omega+0.5q\}$ bodies appears, whose radial cross sections are correspondingly four and three angular plane figures, i.e. for each natural $\omega = 0, 1, 2, \ldots$*

$$GML_4^{4\omega+q} \xrightarrow{VS_{0,2}} link-2\ of\ the\ GML_4^{(4\omega+2)}\{0\}\ and\ GML_3^{6\omega+6}\left\{\omega+\frac{1}{2}\right\} \tag{2.2.16}$$

⋆ ***Case D.*** *after a $V_{0,2}$-zone-slit of the GML_4^n body, an object bulk link-1 (or knot), of the $GML_3^{3\cdot2(\omega+1)}\{\omega+0.5q\}$ body appears. Its radial cross section is given by three angular plane figures, i.e. for each natural $\omega = 0, 1, 2, \ldots$*

$$GML_4^{4\omega+q} \xrightarrow{V_{0,2}} link-1\ of\ the\ GML_3^{6\omega+6}\left\{\omega+\frac{1}{2}\right\} \tag{2.2.17}$$

Sketch of the proof. For the proof the previous argument can be repeated verbatim (see Theorems 3 and 4), but it is necessary to use the corresponding results of Theorem 2., when the greatest common divisor of $m = 4$ and $q = 2$ is equal to 2!

As can be seen from Theorem 2, there are two fundamentally different cases:

⋆ 1. When the s-line does not contain the center of symmetry of the radial cross section of the initial figure (before cutting, Theorem 2. case A (2.1.3)) - corresponding cases for GML_4^n body are (cases B.I. and C.);

⋆ 2. The opposite case b-line contains this point (Theorem 2. case B (2.1.4) - corresponding cases for body are (cases B.II. and D.).

Corollary 3 *In the case of GML bodies in addition there are actual distinctions of shapes of plane elements of the radial cross section of figures which appear after cutting. The form of "new figures" (i.e. figures which appear after the cutting process) depends on the location of the cut surface (more precisely the location of the corresponding line in the radial cross section)!*

⋆ *The new effect, which appeared in the previous Theorem 3.2, in this case is lost, i.e. after cutting process never more than two objects appear!*

⋆ *The Möbius phenomenon is preserved - i.e. after a full cut appears only one object (see cases B.II. and D.)!*

All possible cases and corresponding shapes of the radial cross sections after the cutting process are given in the table in Fig. 6.

2.3 *Relations Between the Set of Generalized Möbius–Listing's Bodies $GML_5^n\{\nu\}$ and the Sets of Bulky Knots and Bulky Links*

In this part similar to the previous part, without loss of generality, and to simplify the proofs of theorems, we will consider initial (before cutting) Generalized Möbius–Listing's Body, whose radial cross section is a **regular pentagon** (corresponding parameters in Gielis Superformula for (1.1.3) function $r(\tau, \theta)$) and its basic line (see Definition 1.3) is a circle. This means that the analytic representation (2.0.2) of corresponding body is more simple, and has the following form

$$\begin{cases} X(\tau,\theta) = \left(R + r(\tau,\theta)\cos\left(\psi + \dfrac{n\theta}{5}\right)\right)\cos(\theta)), \\ Y(\tau,\theta) = \left(R + r(\tau,\theta)\cos\left(\psi + \dfrac{n\theta}{5}\right)\right)\sin(\theta)), \\ Z(\tau,\theta) = r(\tau,\theta)\sin\left(\psi + \dfrac{n\theta}{5}\right), \end{cases} \tag{2.3.1}$$

Using the technique described in [1, 3, 13], we obtain the following theorems:

Theorem 6 *If the number of twisting is $n = 5\omega$, where $\omega \in \mathbb{Z}$ (ω is the number of full rotations of the radial cross section of a body around its basic line) and the GML_5^n body is cut along some of its slit-surfaces, then always an object "bulk link-2" $\{(2\omega)_1^2\}$ (according to the classic tabulation of the links of small complexity*

Shapes of radial cross sections $GML_4^{4\omega+2}$	-slit	Parameters of the objects which appear after cutting		
		Shapes of radial cross sections	Structure of elements	Link group of object
A. 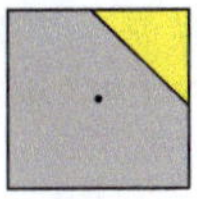	$S_{1,2}$	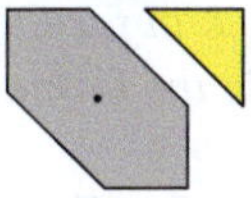	$GML_6^{6\omega+3}$ $GML_3^{6\omega+6}\{\omega+0.5\}$ 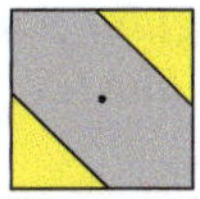	Link-2
B.I 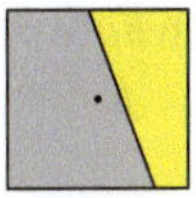	$S_{1,2}$	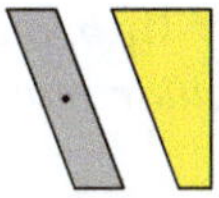	$GML_4^{4\omega+2}$ $GML_4^{8\omega+8}\{\omega+0.5\}$ 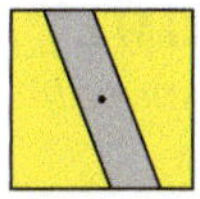	Link-2
B.II 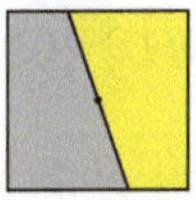	SB	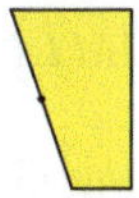	$GML_4^{8\omega+8}\{\omega+0.5\}$ 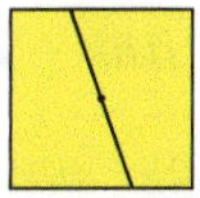	Link-1
C. 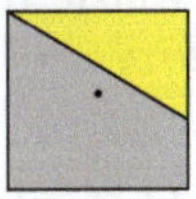	$VS_{0,2}$	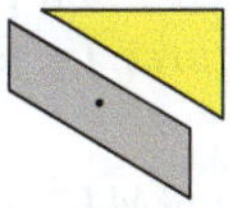	$GML_4^{4\omega+2}$ $GML_3^{6\omega+6}\{\omega+0.5\}$ 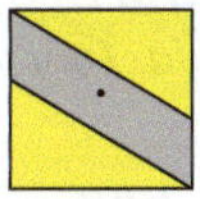	Link-2
D. 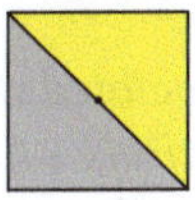	$V_{0,2}$	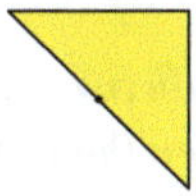	$GML_3^{6\omega+6}\{\omega+0.5\}$ 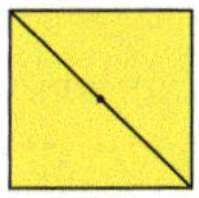	Link-1

Fig. 6 Table of all possible cases and corresponding shapes of the radial cross sections which appear after the cutting process of GML bodies, with symmetry number $m = 4$ and even number of rotations, but not multiple of 4

[9–11]) of the two bulk link-1 appears; But both components of this bulk link-2 have ***five different geometric structures****, more precisely:*

⋆ ***Case A.*** *- after an $S_{1,2}$-zone-slit of the GML_5^n body, an object bulk link-2 $\{(2\omega)_1^2\}$ of the $GML_3^{3\omega}\{\omega\}$ and $GML_6^{6\omega}\{0\}$ bodies, appears whose radial cross sections are correspondingly 3 and 6 angular plane figures, i.e. for each natural $\omega = 0, 1, 2, \ldots$*

$$GML_5^{5\omega} \xrightarrow{S_{1,2}} link - 2\{(2\omega)_1^2\}\ of\ the\ GML_3^{3\omega}\{\omega\}\ and\ GML_6^{6\omega}\{0\} \qquad (2.3.2)$$

⋆ ***Case B.*** *- after an $S_{1,3}$-zone-slit of the GML_5^n body, an object bulk link-2 $\{(2\omega)_1^2\}$ of the $GML_4^{4\omega}\{\mu\}$ and $GML_5^{5\omega}\{\nu\}$ bodies, appears whose radial cross sections are four and five angular plane figures, i.e. for each natural $\omega = 0, 1, 2, \ldots$*

$$GML_5^{5\omega} \xrightarrow{S_{1,3}} link - 2\{(2\omega)_1^2\}\ of\ the\ GML_4^{4\omega}\{\mu\}\ and\ GML_5^{5\omega}\{\nu\} \qquad (2.3.3)$$

There are three fundamentally different subcases, that are associated with finding the center of symmetry of initial figures after the cut; more precisely - if the center of symmetry of the radial cross section (initial pentagon) after cutting remains in:

⋆ *a. the four-angular part of domain, then in formula (2.3.3)* $\mu = 0$ *and* $\nu = \omega$;

⋆ *b. the five-angular part of domain, then in formula (2.3.3)* $\mu = \omega$ *and* $\nu = 0$;

⋆ *c. the cutting line (i.e.* SB*-slit), then in the formula both elements have characteristic* $\mu = \nu = \omega$.

Case C.I. *- after an* $VS_{0,2}$*-zone-slit of the* GML_5^n *body, an object bulk link-2* $\{(2\omega)_1^2\}$ *of the* $GML_3^{3\omega}\{\omega\}$ *and* $GML_5^{5\omega}\{0\}$ *bodies appears, whose radial cross section are correspondingly 3 and 5 angular plane figures, i.e. for each natural* $\omega = 0, 1, 2, \ldots$

$$GML_5^{5\omega} \overset{VS_{0,2}}{\longrightarrow} link - 2\{(2\omega)_1^2\} \; of \; the \; GML_3^{3\omega}\{\omega\} \; and \; GML_5^{5\omega}\{0\} \qquad (2.3.4)$$

Case C.II. *- after an* $VS_{0,3}$*-zone-slit of the* GML_5^n *body, an object bulk link-2* $\{(2\omega)_1^2\}$ *of the* $GML_4^{4\omega}\{\mu\}$ *and* $GML_4^{4\omega}\{\nu\}$ *bodies appears, whose radial cross section are 4 angular (not identic) plane figures, i.e. for each natural* $\omega = 0, 1, 2, \ldots$

$$GML_5^{5\omega} \overset{VS_{0,3}}{\longrightarrow} link - 2\{(2\omega)_1^2\} \; of \; the \; GML_4^{4\omega}\{\mu\} \; and \; GML_4^{4\omega}\{\nu\} \qquad (2.3.5)$$

There are two fundamentally different subcases, that are associated with finding the center of symmetry of initial figures after the cut; more precisely - if the center of symmetry of the radial cross section (initial pentagon) after cutting remains in:

⋆ *a. one four-angular part of domain, then that figure, after the cut, has a characteristic* $\nu = 0$*, i.e. in formula (2.3.5)* $\mu = \omega$ *and* $\nu = 0$;

⋆ *b. the cutting line (i.e.* $VBS_{0,3}$*-slit), then in formula (2.3.5) both element have characteristic* $\mu = \nu = \omega$*. In this case both geometric objects are identical.*

Case D. *- after an* $V_{0,2}$*-zone-slit of the* GML_5^n *body, an object bulk link-2* $\{(2\omega)_1^2\}$ *of the* $GML_3^{3\omega}\{\omega\}$ *and* $GML_4^{4\omega}\{0\}$ *bodies appears, whose radial cross section are correspondingly 3 and 4 angular plane figures, i.e. for each natural* $\omega = 0, 1, 2, \ldots$

$$GML_5^{5\omega} \overset{V_{0,2}}{\longrightarrow} link - 2\{(2\omega)_1^2\} \; of \; the \; GML_3^{3\omega}\{\omega\} \; GML_4^{4\omega}\{0\} \qquad (2.3.6)$$

Sketch of the proof. We draw straight lines connecting the center of symmetry with the corresponding vertices, so we obtain a star like cross, with five wings, and this case of generalized Möbius–Listing's surfaces (GML_5^n for all numbers of m and n), has already been studied (see Theorem 1, and full text of proof e.g. [1, 2]). Slit surface for bulky link is converted in a corresponding slit-line, but in this situation we know all possible variants that appear after cutting. As can be seen from Theorem 1, there are two fundamentally different cases:

⋆ 1. When the s-line does not contain the center of symmetry of the radial cross section of the initial figure (before cutting, Theorem 1, case A (2.1.1)) - corresponding cases for GML_5^n body are (cases A, B.a, B.b, C.I, C.IIa and D in Theorem 6);

⋆ 2. In the opposite case, the b-line contains this point (Theorem 1. case B (2.1.2)) - corresponding cases for GML_5^n body are (cases B.c and C.II.b).

In the case of GML bodies in addition there are actual distinctions of shapes of plane elements of the radial cross section of figures which appear after cutting. The form of "new figures" (i.e. figures which appear after the cutting process) depends on the location of the cut surface (more precisely the location of the corresponding line in the radial cross section)!

⋆ The new effect, which appeared in the previous Theorem 4, in this case is lost, i.e. after cutting process never appear more then two objects!

⋆ In this case, also the Möbius phenomenon is lost - i.e. after a full cut never appears only one object, but it is more or less natural!

But now we return to the considering case and we count separately how many rotations make some vertices of new bodies that appear after cutting! All possible cases and corresponding shapes of the radial cross sections after the cutting process are given in the table in Fig. 7.

Theorem 7 *If the number of twisting is $n = 5 \cdot \omega + q$, $q = 1, 2, 3, 4$, where $\omega \in \mathbb{Z}$ and the GML_5^n body is cut along some of its slit-surfaces, then* ***twelve different cases*** *from the geometric forms of the radial cross sections appear. More precisely the following results hold:*

⋆ ***Case A.*** *Taking into account Remarks 2.1 and 2.2, we have three different subcases:*

∘ ***Subcase A.I.*** *if $b_2^1 + b_2^2 < A$; (see Remark 2.2-A.1), after an $S_{1,2}$-zone-slit of the GML_5^n body, an object bulk link-2, of the $GML_{10}^{2(5\omega+q)}\{0\}$ and $GML_3^{3(5(\omega+1)+(q-1))}$ $\{\omega + 0.2q\}$ bodies appear. The first one has the same structure $\{(0)_1\}$ as the initial body before cutting. Also, their radial cross sections are ten and three angular plane figures (usually there are not regular polygons), i.e. for each natural $\omega = 0, 1, 2, \ldots$*

$$GML_5^{5\omega+q} \xrightarrow{S_{1,2}} link-2\ of\ the\ GML_{10}^{10\omega+2q)}\{0\}\ and\ GML_3^{15\omega+3q+12}\left\{\omega+\frac{q}{5}\right\} \tag{2.3.7}$$

∘ ***Subcase A.II.*** *if $b_2^1 + b_2^2 = A$; (see Remark 2.2-A.2), after an $S_{1,2}$-zone-slit of the GML_5^n body, an object bulk link-2, of the $GML_5^{(5\omega+q)}\{0\}$ and $GML_3^{3(5(\omega+1)+(q-1))}$ $\{\omega + 0.2q\}$ bodies appear. The first one has the same structure $\{(0)_1\}$ as the initial body before cutting. Also their radial cross sections are five and three angular plane figures, i.e. for each natural $\omega = 0, 1, 2, \ldots$*

$$GML_5^{5\omega+q} \xrightarrow{S_{1,2}} link-2\ of\ the\ GML_5^{(5\omega+q)}\{0\}\ and\ GML_3^{15\omega+3q+12}\left\{\omega+\frac{q}{5}\right\} \tag{2.3.8}$$

∘ ***Subcase A.III.*** *if $b_2^1 + b_2^2 > A$; (see Remark 2.2-A.3), after an $S_{1,2}$-zone-slit of the GML_5^n body, an object bulk link- 3, of the $GML_5^{(5\omega+q)}\{0\}$, $GML_3^{3(5(\omega+1)+(q-1))}$ $\{\omega + 0.2q\}$ and $GML_5^{5(5(\omega+1)+(q-1))}\{\omega + 0.2q\}$ bodies appear. The first one has the*

Shapes of radial cross section $GML_5^{5\omega}$	-slit	Parameters of the objects which appear after cutting: Shapes of radial cross sections	Structure of elements	Link group of object
A. 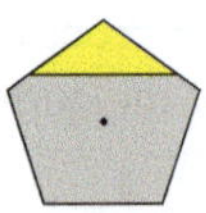	$S_{1,2}$	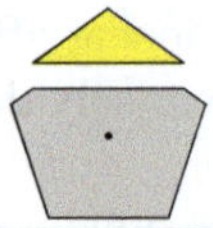	$GML_3^{3\omega}\{\omega\}$ $GML_6^{6\omega}$	
B. a. 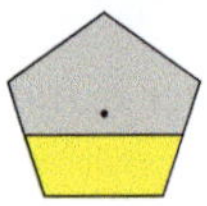	$S_{1,3}$	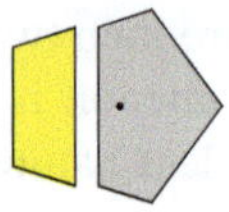	$GML_4^{4\omega}\{\omega\}$ $GML_5^{5\omega}$	
b. 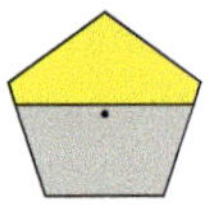	or	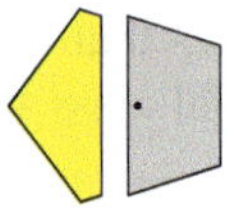	$GML_4^{4\omega}$ $GML_5^{5\omega}\{\omega\}$	
c. 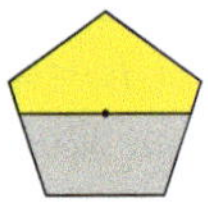	SB	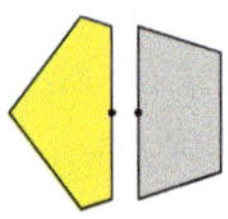	$GML_4^{4\omega}\{\omega\}$ $GML_5^{5\omega}\{\omega\}$	Link-2 $\{(2\omega)_1^2\}$
C. I. 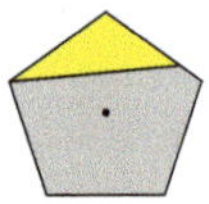	$VS_{0,2}$	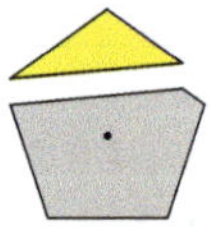	$GML_3^{3\omega}\{\omega\}$ $GML_5^{5\omega}$	
II. a. 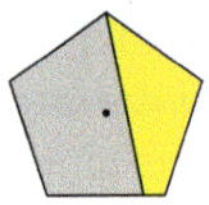	$VS_{0,3}$	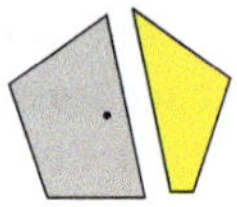	$GML_4^{4\omega}\{\omega\}$ $GML_4^{4\omega}$	
II. b. 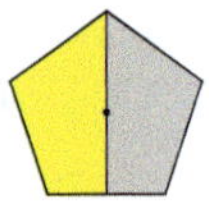	VBS	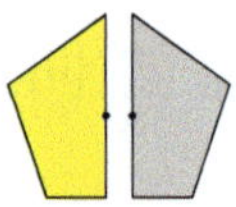	$GML_4^{4\omega}\{\omega\}$ $GML_4^{4\omega}\{\omega\}$	
D. 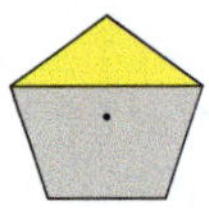	$V_{0,2}$	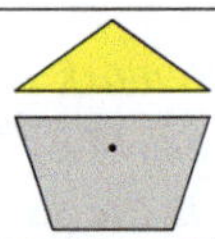	$GML_3^{3\omega}\{\omega\}$ $GML_4^{4\omega}$	

Fig. 7 Table of all possible cases and corresponding shapes of the radial cross sections which appear after the cutting process of GML bodies, with symmetry number $m = 5$ and number of rotations multiples of 5

same structure $\{(0)_1\}$ *as the initial body before cutting. Also their radial cross sections are correspondingly five, three and five angular plane figures, i.e. for each natural* $\omega = 0, 1, 2, \ldots$

$$GML_5^{5\omega+q} \xrightarrow{S_{1,2}} link-3\ of\ the\ GML_5^{(5\omega+q)}\{0\}\,, \\ GML_3^{15\omega+3q+12}\left\{\omega+\frac{q}{5}\right\}\ and\ GML_5^{25\omega+5q+20}\left\{\omega+\frac{q}{5}\right\} \tag{2.3.9}$$

⋆ ***Case B.*** *After an $S_{1,3}$-zone-slit (which is not a SB-zone-slit (see Remark 2.2-B.5 and Definition 2.5), we have four different subcases:*

∘ ***Subcase B.I.*** *if $b_3^1 + b_3^2 < A$ (see Remark 2.2-B 1), after an $S_{1,3}$-zone-slit of the GML_5^n body, an object bulk link-3, of the $GML_5^{(5\omega+q)}\{0\}$ and two $GML_4^{4(5(\omega+1)+(q-1))}\{\omega+0.2q\}$ bodies appear. The first one has the same structure $\{(0)_1\}$ as the initial body before cutting. Also their radial cross sections are correspondingly five and four (not identic) angular plane figures, i.e. for each natural $\omega = 0, 1, 2, \ldots$*

$$GML_5^{5\omega+q} \xrightarrow{S_{1,3}} link-3\ of\ the\ GML_5^{(5\omega+q)}\{0\}\,,\ and\ two\ GML_4^{20\omega+4q+16}\left\{\omega+\frac{q}{5}\right\} \tag{2.3.10}$$

∘ ***Subcase B.II.*** *if $b_3^1 + b_3^2 = A$ (see Remark 2.2-B 2), after an $S_{1,3}$-zone-slit of the GML_5^n body an object bulk link-3, of the $GML_5^{(5\omega+q)}\{0\}$, $GML_3^{3(5(\omega+1)+(q-1))}\{\omega+0.2q\}$ and $GML_4^{4(5(\omega+1)+(q-1))}\{\omega+0.2q\}$ bodies appear. The first one has the same structure $\{(0)_1\}$ as the initial body before cutting. Also their radial cross sections are correspondingly five, three and four angular plane figures, i.e. for each natural $\omega = 0, 1, 2, \ldots$*

$$GML_5^{5\omega+q} \xrightarrow{S_{1,3}} link-3\ of\ the\ GML_5^{(5\omega+q)}\{0\}\,, \\ GML_3^{15\omega+3q+12}\left\{\omega+\frac{q}{5}\right\}\ and\ GML_4^{20\omega+4q+16}\left\{\omega+\frac{q}{5}\right\} \tag{2.3.11}$$

∘ ***Subcase B.III.*** *if $b_3^1 + b_3^2 > A$ and one of the $b_3^i < 0.5A$, where $i = 1, 2$ (see Remark 2.2-B 3), after an $S_{1,3}$-zone-slit of the GML_5^n body an object bulk link-4, of the $GML_5^{(5\omega+q)}\{0\}$, two similar bodies (but not with identical radial cross sections) $GML_3^{3(5(\omega+1)+(q-1))}\{\omega+0.2q\}$ and $GML_6^{6(5(\omega+1)+(q-1))}\{\omega+0.2q\}$ bodies appear. Only the first one has the same structure $\{(0)_1\}$ as the initial body before cutting. Also their radial cross sections are correspondingly five, three, three and six angular plane figures, i.e. for each natural $\omega = 0, 1, 2, \ldots$*

$$GML_5^{5\omega+q} \xrightarrow{S_{1,3}} link-4\ of\ the\ GML_5^{(5\omega+q)}\{0\}\,, \\ two\ GML_3^{15\omega+3q+12}\left\{\omega+\frac{q}{5}\right\}\ and\ GML_6^{30\omega+6q+24}\left\{\omega+\frac{q}{5}\right\} \tag{2.3.12}$$

∘ ***Subcase B.IV.*** *if $b_3^1 + b_3^2 > A$ and both of the $b_3^i > 0.5A$, where $i = 1, 2$ (see Remark 2.2-B 4), after an $S_{1,3}$-zone-slit of the GML_5^n body an object bulk link-4, of the $GML_5^{(5\omega+q)}\{0\}$, $GML_3^{3(5(\omega+1)+(q-1))}\{\omega+0.2q\}$, $GML_4^{4(5(\omega+1)+(q-1))}\{\omega+0.2q\}$ and $GML_5^{5(5(\omega+1)+(q-1))}\{\omega+0.2q\}$ bodies appear. Only the first one has the same* *structure* $\{(0)_1\}$ *as the initial body before cutting. Also their radial cross sections are correspondingly five, three, four and five angular plane figures, i.e. for each natural* $\omega = 0, 1, 2, \ldots$

$$GML_5^{5\omega+q} \xrightarrow{S_{1,3}} link-4\ of\ the\ GML_5^{(5\omega+q)}\{0\},\ GML_3^{15\omega+3q+12}\left\{\omega+\frac{q}{5}\right\},$$
$$GML_4^{20\omega+4q+16}\left\{\omega+\frac{q}{5}\right\}\ and\ GML_5^{25\omega+5q+20}\left\{\omega+\frac{q}{5}\right\} \tag{2.3.13}$$

⋆ ***Case C.*** *after an* SB*-zone-slit of the* GML_5^n *body (see Remark 2.2-B 5), an object bulk link-2 of the* $GML_3^{3(5(\omega+1)+(q-1))}\{\omega+0.2q\}$ *and* $GML_4^{4(5(\omega+1)+(q-1))}\{\omega+0.2q\}$ *body appear. Their radial cross section are three and four angular plane figures. Only the first one has the same structure* $\{(0)_1\}$ *as the initial body before cutting; i.e. for each natural* $\omega = 0, 1, 2, \ldots$

$$GML_5^{5\omega+q} \xrightarrow{SB} link-1\ of\ the\ GML_3^{15\omega+3q+12}\left\{\omega+\frac{q}{5}\right\}\ and\ GML_4^{20\omega+4q+16}\left\{\omega+\frac{q}{5}\right\} \tag{2.3.14}$$

⋆ ***Case D.*** *after an* $VS_{0,2}$*-zone-slit of the* GML_5^n *body, an object bulk link-3, of the* $GML_5^{(5\omega+q)}\{0\}$*,* $GML_3^{3(5(\omega+1)+(q-1))}\{\omega+0.2q\}$ *and* $GML_4^{4(5(\omega+1)+(q-1))}\{\omega+0.25q\}$ *bodies appear. Only the first one has the same structure* $\{(0)_1\}$ *as the initial body before cutting, whose radial cross sections are correspondingly five, three and four angular plane figures, i.e. for each natural* $\omega = 0, 1, 2, \ldots$

$$\begin{aligned} &GML_5^{5\omega+q} \xrightarrow{VS_{0,2}} link-3\ of\ the\ GML_5^{(5\omega+q)}\{0\},\\ &GML_3^{15\omega+3q+12}\left\{\omega+\frac{q}{5}\right\}\ and\ GML_4^{20\omega+4q+16}\left\{\omega+\frac{q}{5}\right\}\end{aligned} \tag{2.3.15}$$

⋆ ***Case E.*** *after an* $VS_{0,3}$*-zone-slit of the* GML_5^n *body, an object bulk link-4, of the* $GML_5^{(5\omega+q)}\{0\}$*, two bodies* $GML_3^{3(5(\omega+1)+(q-1))}\{\omega+0.2q\}$ *and* $GML_5^{5(5(\omega+1)+(q-1))}$ $\{\omega+0.2q\}$ *bodies appear; Only the first one has the same structure* $\{(0)_1\}$ *as the initial body before cutting; Their radial cross sections are correspondingly five, three, three and five angular plane figures, i.e. for each natural* $\omega = 0, 1, 2, \ldots$

$$\begin{aligned} &GML_5^{5\omega+q} \xrightarrow{VS_{0,3}} link-4\ of\ the\ GML_5^{(5\omega+q)}\{0\},\\ &two\ GML_3^{15\omega+3q+12}\left\{\omega+\frac{q}{5}\right\}\ and\ GML_5^{25\omega+5q+20}\left\{\omega+\frac{q}{5}\right\}\end{aligned} \tag{2.3.16}$$

⋆ ***Case F.*** *after an* $VBS_{0,3}$*-zone-slit of the* GML_5^n *body, an object bulk link-2, of the two* $GML_3^{3(5(\omega+1)+(q-1))}\{\omega+0.2q\}$ *bodies appear, whose radial cross sections are three angular (not identic) plane figures, i.e. for each natural* $\omega = 0, 1, 2, \ldots$

$$GML_5^{5\omega+q} \xrightarrow{VBS_{0,3}} link-2\ of\ the\ two\ GML_3^{15\omega+3q+12}\left\{\omega+\frac{q}{5}\right\} \tag{2.3.17}$$

★ ***Case G.*** *after an $V_{0,2}$-zone-slit of the GML_5^n body, an object bulk link-3, of the $GML_5^{(5\omega+q)}\{0\}$ and two $GML_3^{3(5(\omega+1)+(q-1))}\{\omega+0.2q\}$ body appears. Its radial cross section are correspondingly five and three angular plane figures and only the first one has the same structure $\{(0)_1\}$ as the initial body before cutting; i.e. for each natural $\omega = 0, 1, 2, \ldots$*

$$GML_5^{5\omega+q} \xrightarrow{V_{0,2}} link-3\ of\ the\ GML_5^{(5\omega+q)}\{0\}\ and\ two\ GML_3^{15\omega+3q+12}\left\{\omega+\frac{q}{5}\right\} \tag{2.3.18}$$

Sketch of the proof. For the proof we can repeat verbatim the previous argument (Theorem 6), but it is necessary to use the corresponding results of Theorem 2, when the greatest common divisor of $m = 5$ and q is equal to 1!

As can be seen from Theorem 2, there are two fundamentally different cases:

★ 1. When the s-line does not contain the center of symmetry of the radial cross section of the initial figure (before cutting, Theorem 2. case A (2.1.3)) - corresponding cases for GML_5^n body are (cases A.I., A.II., A.III., B.I., B.II., B.III., B.IV., D., E. and G.);

★ 2. In the opposite case, the b-line contains this point (Theorem 2. case B (2.1.4)) - corresponding cases for body are (cases C. and F.).

Corollary 4 *In the case of GML bodies in addition there are actual distinctions of shapes of plane elements of the radial cross section of figures which appear after cutting. The form of "new figures" (i.e. figures which appear after the cutting process) depends on the location of the cut surface (more precisely the location of the corresponding line in the radial cross section)!*

★***A new effect is that the s-surface can be self-intersecting, and therefore appears in more than two objects*** *(see cases A.III., B.I., B.II., B.III., B.IV., D., E. and G.)!*

★ *But in this case (when the symmetry number is 5) the Möbius phenomenon is not preserved! - i.e. after a full cut appears only one object!*

Finally, we return to the considering case and we count separately how many rotation make some vertices of new bodies that appear after cutting!

All possible cases and corresponding shapes of the radial cross sections after the cutting process are given in the table in Fig. 8.

2.4 Similarity and Difference. Relations Between the Sets of Generalized Möbius–Listing's Bodies and Surfaces With Different Numbers of Symmetry

We return to the six questions formulated at the beginning of this article, and we try to answer individually and to make some generalizations (see examples in Fig. 9).

Answers to the first three questions are given correspondingly in Theorems 1–7. This shows that while solutions for GML_m^n surfaces are given for all cases, for any

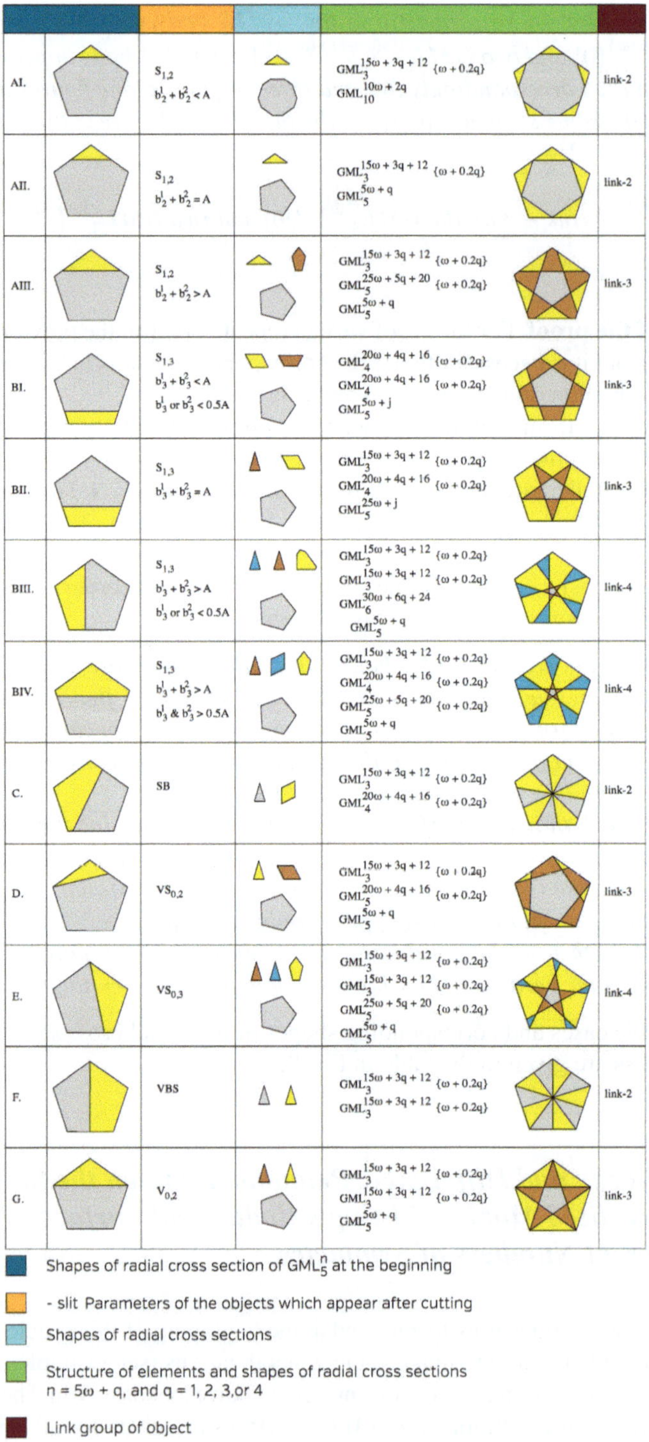

Fig. 8 (Continued)

◀**Fig. 8** Table of all possible cases and corresponding shapes of the radial cross sections which appear after the cutting process of GML, with symmetry number $m = 5$ and number of rotations not multiple of 5

numbers m (of symmetry of the radial cross section) and n (of rotation of the ends of corresponding prism before identifying), instead similar problems for GML_m^n bodies are investigated, at this stage, only for some concrete values of the number m ($m = 2, 3, 4, 5$ - see [4–7]).

Remark 2.3 Similarity and difference between corresponding ribbon and bulky links ($m = 3, 4$ or 5).

• A. - Similarity. If $n = m\omega$, then groups of bulk links which appear after cutting of the GML_m^n body are similar to the ribbon links which appear after an s-zone slit of the GML_m^n surface (see [1]); on these similarities the proofs of Theorems 3 and 6 were based;

• B. - Differences. If $n = m\omega$ then, after cutting of a GML_m^n body never appears bulk link which is analogous to the ribbon links which appear after a b-zone slit of the surface (see Theorem 1. B. opposite of Theorems 3 and 6; see also [4–7]);

⋆ - These are **new phenomena and exist only** for GML_m^n surfaces with star like plane figures in the radial cross sections - after one cutting appears object with more than two independent components (i.e. link-j and $2 < j \leq m$).

• C. - Similarity. If $n = m\omega + q$, when $\gcd(m, q) = 1$, then set of the bulk link -1 which appears after cutting of a GML_m^n bodies is similar to the ribbon link -1 which appears after an s-zone slit of the corresponding surfaces, i.e. these objects individually have similar characteristics of the basic line $\{\omega + q/m\}$; on these similarities the proofs of Theorems 4 and 7 were based (see also [1] or [2]);

• D. - Differences. For arbitrary n after an s-zone slit of a GML_m^n surface never appears a geometric object with more than two independent components, i.e. ribbon link-3 or link-4 with three or four components (similar cases: A.III., B.I., C. of Theorem 4 and A.III., B.I. - B.IV., D., E. and G. of Theorem 7).

• E. - Differences. For arbitrary n after an s-zone slit of the GML_m^n surface never appears ribbon link-j ($j > 2$) with three or more components. These are **new phenomena and exist only for GML_m^n bodies**, when after one cutting an object

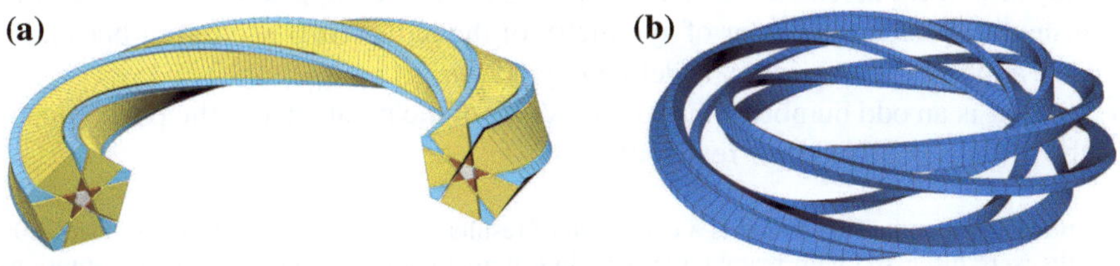

Fig. 9 Examples of GML bodies which appear after the cutting process. a) a geometrical structure with 4 independent objects; b) one object—repetition of the classic phenomenon of Möbius strip

with more than two independent components appears; unfortunately we have not yet found the general law for computing the full number of the different cases, for arbitrary numbers of m and n, more precisely we have some results but at present we have no general proof.

• F. - Similarity. For arbitrary numbers m and n after an s-zone slit always one component of the arising link is similar to the initial GML_m^n body or surface.

• G. - Similarity. For arbitrary numbers m and n after a b-zone slit of GML_m^n bodies or surfaces the initial body structure disappears, i.e. characteristics of the arising objects is never equal to zero ($\{\omega + q/m\}$, $q = 0, 1, 2, \dots, m-1$)!

• H. - Differences. For arbitrary numbers m and n after a b-zone slit of GML_m^n bodies the number of independent components of the arising object is never more than two! But, for arbitrary numbers m and n after a b-zone slit of GML_m^n surfaces, the number of independent components of the arising object may be more than two. More precisely, this number equals the numbers m or $\varsigma \equiv \gcd(m, q)$ (see correspondingly in Theorems 1 or 2, case B)!

Remark 2.4 Generalisations.

• A. - If m is an even number, then for different n (more precisely, if $\gcd(m, n) = 1$) - after one full cutting of GML_m^n bodies, maximum $\dfrac{m}{2} + 1$ independent geometric objects appear (this number depends also on the geometric place of the cutting line in the cross section of body), i.e. link-$(m/2 + 1)$ appear and only one element has structure similar to figure before cutting;

• B. - If m is an odd number, then for different n (more precisely, if $\gcd(m, n) = 1$) - after one full cutting of GML_m^n bodies, maximum $\left[\dfrac{m}{2}\right] + 2$ independent geometric objects appear (this number depends also on the geometric place of the cutting line in the cross section of body), i.e. link-$([m/2] + 2)$ appear and only one element has structure similar to figure before cutting;

• C. - If m is an even number, then always some values of n exist, such that after one full cutting of GML_m^n bodies only 1 independent geometric object appears (for this, the cutting line should include the center of symmetry of the radial cross section of the body), i.e. knot (link-1 appears, whose index is defined by $\gcd(m, n) = 1$;

⋆ - If m is an even number, then always some values of n exist, and the phenomenon of the Möbius strip is realized!

• D. - If m is an odd number, then there exist some values of n, such that after one cutting of GML_m^n at least 2 independent geometric objects appear (for this the cutting line should include the center of symmetry of the radial cross section of body), i.e. (link-2 appears), whose index is defined by $\gcd(m, n) = 1$;

⋆ - If m is an odd number, then for any value of the parameter n, the phenomenon of the Möbius band is never realized!

Acknowledgements This article is a collection of results obtained at different times since 2006. Results were discussed with many colleagues in different countries and in different institutions. For the authors were very valuable discussions with professors P.V. Checcerini and M. Tallini Scafati in 2008 in Rome. A special role is played by creative contact with Johan Gielis, with which some important results have been obtained. At different times and different results were obtained with C. Cassisa and D. Caratelli; For valuable discussions, the authors thank colleagues N. Avazashvili, M. Nakashidze, P.P. Natalini, S. Camiz. Authors are grateful to Ph.D. Student of St. Andrew the First-Called Georgian University of the Patriarchate of Georgia Levan Roinishvili for creating the convenient program for demonstration of the obtained mathematical results. With special gratitude, we want to mark the valuable assistance of Albert Kiefer on graphics. Some part of results has been fulfilled by a financial support of Shota Rustaveli National Science Foundation (Grant SRNSF/FR/358/5-109/14).

References

1. I. Tavkhelidze, About connection of the generalized Möbius Listing's surfaces with sets of Ribbon Knots and links, in *Proceedings of Ukrainian Mathematical Congress, S.2 Topology and Geometry* (Kiev, 2011), pp. 117–129
2. I. Tavkhelidze, C. Cassisa, P.E. Ricci, About connection of the generalized Möbius Listing's surfaces with sets of knots and links. Lecture Notes of Seminario Interdisciplinare di Matematica **9**, 187–200 (2010)
3. J. Gielis, D. Caratelli, Y. Fougerolle, P.E. Ricci, I. Tavkhelidze, T. Gerats, Universal natural shapes from unifying shape description to simple methods for shape analysis and boundary value problems. PLoSOne (2012). doi:10.1371/journal.pone.0029324
4. I. Tavkhelidze, About structure and some geometric characteristic of the bulk links which appear after cutting of generalized Möbius-Listings bodies, in *Proceedings of I. Vekua Institute of Applied Mathematics*, vol. 65 (2015)
5. I. Tavkhelidze, C. Cassisa, J. Gielis, P.E. Ricci, About "Bulky" links, generated by generalized Möbius-Listing's bodies GML_3^n. Rend. Lincei Mat. Appl. **24**, 11–38 (2013)
6. D. Caratelli, J. Gielis, P.E. Ricci, I. Tavkhelidze, Some properties of bulky links generated by generalized Möbius-Listing's bodies GML_4^n. J. Math. Sci. **216**(4), 509–518 (2016)
7. D. Caratelli, J. Gielis, P.E. Ricci, I. Tavkhelidze, Some properties of "Bulky" links, generated by generalized Möbius Listing's bodies GML_4^n. in *Proceedings of the International Conference - Lie Groups, Differential Equations and Geometry*, June 10–22 (Batumi, Georgia, 2013), pp. 30–43
8. A. Sosinsky, *Knots: mathematics with a Twist* (Harvard University Press, Cambridge, 2002), pp. 1–147
9. E.W. Weisstein, *The CRC concise Encyclopedia of Mathematic*, 2nd edn. (CRC Press, Florida, 2001)
10. H. Doll, J. Hoste, A tabulation of oriented links. Math. Comput. **57**, 747–761 (1991)
11. G. Kupenberg, Quadrisecants of knots and links. J. Knot Theory Ramifications **3**, 41–50 (1994)
12. A. Gamkrelidze, Algorithms in low-dimensional topology: holonomic parametrization of knots. J. Math. Sci. **195**(2), (2013)
13. I. Tavkhelidze, P.E. Ricci, Classification of a wide set of geometric figures, surfaces and lines (Trajectories), Rendiconti Accademia Nazionale delle Scienze detta dei XL, Memorie di Matematica e Applicazioni, 124°, vol. XXX, fasc. 1, pp. 191–212 (2006)

MIX
Papier aus verantwortungsvollen Quellen
Paper from responsible sources
FSC® C105338
FSC
www.fsc.org

If you have any concerns about our products,
you can contact us on
ProductSafety@springernature.com

In case Publisher is established outside the EU,
the EU authorized representative is:
Springer Nature Customer Service Center GmbH
Europaplatz 3, 69115 Heidelberg, Germany

Printed by Libri Plureos GmbH
in Hamburg, Germany